W9-DDN-716

The Last Alchemist

The Last Alchemist

COUNT CAGLIOSTRO,

MASTER OF MAGIC IN THE

AGE OF REASON

Iain McCalman

HarperCollins*Publishers*

HarperCollins books may be purchased for educational, business, or sales promotional use. For information, please write: Special Markets Department, HarperCollins Publishers Inc., 10 East 53rd Street, New York, NY 10022.

FIRST EDITION

Designed by Laura Lindgren

Printed on acid-free paper

Library of Congress Cataloging-in-Publication Data
McCalman, Iain.
     The last alchemist : Count Cagliostro, master of magic in the Age of Reason / Iain McCalman.—1st ed.
        p.   cm.
     Includes bibliographical references
     ISBN 0-06-000690-0
     1. Cagliostro, Alessandro, conte di, 1743–1795. 2. Occultists—Europe—Biography.
I. Title.
BF1598.C2M33 2003
133'.092—dc21
[B]   2002191271

03 04 05 06 07 ❖/RRD 10 9 8 7 6 5 4 3 2 1

To David and Verity, my Father and Mother

CONTENTS

LIST OF ILLUSTRATIONS

ACKNOWLEDGMENTS

I have incurred more than my usual quantum of debts in writing this book. It could not have been done without the kindness and expertise of friends who obtained sources for me abroad that I was unable to reach myself, and, on occasion, translated these as well. Natalie Adamson and Serge Kakou deserve special mention here. Others have helped me to translate documents from Italian and German or have checked and improved my own translations in French. I thank them for their generosity and skill. They are Glenn St. John Barclay, Caroline Turner, Bruce Koepke, Susanna von Caemmerer, Mary Caliri, Mike Ovington, and Gino Moliterno. To Gino, special thanks for his enthusiastic support throughout and his expert advice on Cagliostro's career in film. Christa Knellwolf not only translated important documents from the German for me but also gave me the benefit of her expertise on Goethe. Simon Burrows and his wife furnished me with documents I could not otherwise have obtained and Simon gave unstintingly of his expertise on Théveneau de Morande and his milieu. I look forward to his own biography of Théveneau before long. My thanks to John Docker for putting me onto Walter Benjamin's wonderful reflections on Cagliostro.

Georgina Fitzpatrick has my heartfelt thanks for her help with occasional translations, editing the final stages of the book, and searching out permissions for pictures. Having someone so skilled and patient during the birth throes of a book is a tremendous comfort. My thanks, in turn, to the

Australian Research Council for helping to fund such a treasure. Lindy Shultz has also been generous with her skills and advice about layout and picture scanning. Owen Larkin has helped to hunt down potential pictures for the cover. Heather McCalman gave me important early encouragement.

Archivists and librarians have helped me in many ways, especially in the Australian National University; the Australian National Library; the Bibliothèque nationale de France; the Bibliothèque de L'Arsenal; the Archives Nationales; the British Library; the British Museum; the Bodleian Library, Oxford; the Ashmolean Library, Oxford; Torre Abbey, Torquay; the Huntington Library, California; the Archivio de Pesaro; the Staat Archivs in Basel; and the Bibliothèque de la ville de Strasbourg.

To my friend and agent Mary Cunnane, I owe an enormous debt. She gave me the courage and inspiration to attempt the book, and she trained me patiently in how to overcome some academic bad habits. Thanks to my co-agent Peter McGuigan, whose zest, enthusiasm, and advice have proved indispensable at the U.S. end. Dan Conaway of HarperCollins was a wonderful editor; his enthusiasm, patient hard work, and talent have greatly improved the book.

I would also like to thank my friends and acquaintances in Palermo whose warmth and passion for Cagliostro inspired me to write the book. They are Ninetta Cangelosi, Nelida Mendoza, Sebastian, and the Patron of the Ristorante Le Conte de Cagliostro.

Finally, I would like to thank Kate Fullagar, to whom I owe most of all. She has read every draft and given me unending support, encouragement, advice, and insight. I hope this book lives up to your faith.

The Last Alchemist

The House of Balsamo

The greatest enchanter of the eighteenth century deserved better than this, I thought, looking at the house of Count Cagliostro, magician, alchemist, healer, and Freemason. It was a hole bashed high in the side of a crumbling building, halfway along a tiny market lane in Palermo, Sicily. Behind the jagged brick outline you could dimly see the shell of a room, a dusty and derelict cave. The alley stank of urine.

It would be hard to find a sorrier memorial for someone who'd once been a household name throughout the western world, a magician whom monarchs had courted, bishops had feared, artists had painted, doctors had hated, and women had craved. A notice at the entrance told me that the alley's original name, *vicolo delle Perciata*, had been changed to *via conte di Cagliostro*, and it gave a minimal summary of Count Cagliostro's wondrous life.

Giuseppe Balsamo, born in this street in 1743 and educated at the nearby church school of San Rocco, left Palermo at age twenty to visit North Africa, the Levant and the Mediterranean. Adopting the name of Count Cagliostro, he traveled throughout Europe for a decade, performing magical wonders, healing the poor, and founding branches of his Egyptian Masonic movement. Acquitted of swindling after being imprisoned in the Bastille in 1785, he was captured by the Roman Inquisition in 1789 and died in 1795.

His enemies would have been delighted at the neglect of his house. There'd been plenty of them: Casanova, the greatest lover of the age, was bitterly jealous of him; Catherine the Great, empress of Russia, wanted to strangle him; Johann von Goethe, the most revered of Germany's writers, was driven almost mad by hatred of him; King Louis XVI of France persecuted him as a dangerous revolutionary; Queen Marie-Antoinette wanted him locked permanently in the Bastille for involving her in a diamond necklace swindle; and Pope Pius VI accused him of threatening the survival of the Catholic church.

I'd come twelve thousand miles from Australia to see a hole in the wall, the sad remains of the life of Count Cagliostro. What had I hoped to find — a footprint, a ghost, some clue to his mystery? As a historian, I wanted of course to answer the question that's been asked by every one of his previous biographers, and many of his contemporaries as well. Was Count Cagliostro a sinner or a saint — the worst scoundrel of his age or a great occult healer? His biographers, most of them Italian, French, and German, usually came down strongly on one side or the other, but it seemed to me that he might have been both.

His enemies called him a coarse, shallow charlatan, but he'd somehow survived the harsh eraser of history. You can still see Cagliostro today in science fiction movies like *Spawn* or in dramatic costume movies like *The Affair of the Necklace*. At least half a dozen films have been made about his life — in Russia, in America, in Germany, in Italy, and in France. In the last two countries, he still features in television cartoons, comics, pop music, and pulp novels. More highbrow consumers encounter him as Sarastro in Mozart's famous opera *The Magic Flute*, or in Johann Strauss's operetta *Cagliostro in Wien*.

Cagliostro's many attackers also presented another enigma. The Roman Inquisitor Monsignor Barberi, who spent fifteen months interrogating Cagliostro in the Castle of Sant'Angelo, posed it in 1791:

Who could have imagined that a man of his description should have been received with respect in some of the most enlightened cities of Europe? That he should have been regarded as a star propitious in the human race, as a new prophet, and as a type and representation of the Divinity? That he should have approached thrones? That haughty grandees should have become his humble suitors, and nobles paid him the most profound veneration?

A near contemporary, the rabidly brilliant Scottish historian Thomas Carlyle, took Barberi's question a step further by asking, in 1833, how all this could have been achieved by a man devoid of looks, charm, or intellect. Carlyle blamed Cagliostro's success on the eighteenth century itself, which was not, he said, an age of reason and enlightenment, as is usually claimed, but an age of fraud and superstition.

It was easy enough for Carlyle to lampoon Cagliostro as a mediocrity because he never saw the man in the flesh. An equally skeptical Alsatian noblewoman, Madame Henriette Louise d'Oberkirch, who did meet him in 1781, found herself utterly seized by the magician's charisma. His eyes were "indescribable, with supernatural depths—all fire, yet all ice." His voice caressed her, "like a trumpet veiled in crepe." When he came close, "he frightened you and at the same time inspired you with an insatiable curiosity." She could only conclude that "Cagliostro was possessed of a demonic power, he enthralled the mind, paralysed the will."

Whatever his character, I had to admit I was bewitched by the man, and felt a strange affinity with him. Perhaps, as a European who spent my first eighteen years in Africa before migrating to Australia, I was drawn to Balsamo because we shared a bogus African identity. Giuseppe had been born of Moorish-Sicilian stock and raised in one of Palermo's most Arab quarters (once called Al-Gadida). During his years of fame, he'd pretended to be an Egyptian prince and prophet. In a way Cagliostro and I were both African pretenders.

Looking around me in the old Moorish quarter of the Albergheria, I could understand Cagliostro's sense of affiliation with Africa. Goethe, who'd visited the city in 1786 on a quest similar to mine, had observed that Africa seemed to begin here. The Ballaro market that abuts Cagliostro's birthplace looks, feels, and smells like a casbah. It reminded me of parts of Cairo or even of Zanzibar: frying peanut oil, saffron, cloves, garlic, and rotting garbage. The flagstones are streaked with dust blown from North African deserts or smeared with slops tossed from windows and balconies. You have to step carefully because the tenements cast deep shadows. The paint on most of the buildings is covered in fungal-like stains. Bits of iron hold up the door frames; washing flaps on rigging strung between the houses.

It seemed to me that this quarter couldn't have changed much from the days when young Giuseppe Balsamo was the terror of the streets. I found myself wanting badly to know what these tough-looking market people thought about their most famous son — judging from the state of his house, not much.

Yet there were signs of local, if not official, care. Over the entrance of the alley, a crude plywood arch carried the name Cagliostro. Further along, on the ground underneath the mysterious hole, an open concrete framework had been erected to show the shape of his original two-room apartment. Its sides were covered with occult symbols, as well as a blue and ocher mural depicting Cagliostro's world travels. He'd roamed from Palermo to Alexandria, Medina, Rome, Aix, Saint Petersburg, Warsaw, Strasbourg, Paris, London, Switzerland, and on and on. Now and again, he'd doubled back on his tracks, a restless wanderer in search of what — riches, fame, world revolution, spiritual salvation?

While positioning myself between a blob of dog shit and a dead mouse to photograph this simulacrum of Balsamo's house, I met my first local. He was a wiry young man called Sebastian with a dark jaw, faded jeans, a striped T-shirt, and a soft beret. Speaking first in Italian, then in English, he asked if I belonged to a group being taken around the sites of Balsamo's early life by the head of the Associazione di Cagliostro. She wanted to persuade his acting troupe to write and produce a play celebrating Cagliostro's life.

Sebastian was going to need very little persuading to mount the play. He and his actor friends were already gripped by Cagliostro fever. They thought of Cagliostro as a type of Robin Hood figure, a good bandit who plundered the rich and helped the poor. But like most Sicilians, they agreed that he was also a great actor and showman, someone who could make himself all things to all men. It seemed as plausible an interpretation as any.

I listened to the story of Sebastian's life. He'd just returned to Palermo after years of wandering through Brazil, Denmark, Belgium, Britain, and the United States. With no formal training, he'd taught himself to be a street clown, a musician, an actor, a builder, and a raconteur. He was bright, tough, and oddly spiritual, a Palermo adventurer, light on his feet and sharp of wit, using a mixture of charm, opportunism, and intelligence to survive in a bloody world. He was rooted in this place yet completely international, a global traveler willing to try anything. Now he was keen to play the role of Count Cagliostro, the magician who would help transform Palermo from what it had become, a cultural desert ruled over by Mafia crooks. It occurred to me that Sebastian was a latter-day Cagliostro.

Suddenly we remembered Sebastian's appointment with the tour leader. We pushed through the crowded Ballarò market, a carnival of the appetites where Giuseppe Balsamo learned his early repertoire of peasant magic and urban chicanery. Old men with walnut faces offered us beer by the glass; brawny women waved chunks of rose-colored tuna in our faces. Soon we came to a tiny office with a large sign: CAGLIOSTRO ASSOCIAZIONE, COOPERATIVA SOCIALE, PRESIDENT: NINETTA CANGELOSI. It was a shrine to Cagliostro; the walls were hung with icons of the tubby Sicilian magus and his wife, Seraphina — in paper, wood, colored glass, and plastic.

Nina Cangelosi was probably in her early fifties; smartly dressed, with a kind face and sad, dark eyes. She conveyed the effect of *caffè latte*, her brown Sicilian complexion topped with a froth of creamy-silver hair. She gave off the kind of bone weariness that came from throwing oneself continuously against an intractable object. I learned that she'd fought for the last five years to have the Balsamo house restored, lobbying academics, curators, local councillors, government officials, parliamentarians, archivists, businessmen, anyone and everyone, begging them to honor Cagliostro's memory.

Most of them thought that Cagliostro was a crook and therefore not the type whom a modern city should be honoring. Goethe, who has brass plates all over Palermo marking his visit in 1786, met Giuseppe Balsamo's mother and sister, then told the world that Cagliostro was a swindler and a rogue. And Goethe's view has prevailed.

Nina argued with the city worthies that this was not the perception of ordinary Sicilians: they wanted to visit Balsamo's house to see for themselves how humbly this great man began his life. His life story gave them hope. Balsamo had refused to be cowed or trapped by poverty, the curse of Sicily. Nina's enthusiasm had inspired some art students in Palermo to paint the mural, but she found that the city councillors were always too busy to see her. The owner of the building containing the shell of Balsamo's house was gradually buying up most of area with an eye to redevelopment. He was also the patron of the market, a very powerful man. He didn't want a shrine to Cagliostro; he wanted new apartments to sell.

I asked Nina about her passion for Cagliostro. She'd read everything about the magician she could find. She agreed that many bad things had been written about him, mostly taken from the reports of his enemies. Yet even if some accusations were true, she felt that most writers failed to understand the ordeal of making a life in a place as poor as the Albergheria. Perhaps Giuseppe Balsamo was a *mascalzone* (rogue) when young: sadly, she said, many of today's young Palermitans were forced down the same path. But surely no one could deny that Cagliostro atoned for his crimes by healing the poor, free, all over the world and by preaching a message of tolerance and spirituality?

Many people of all nationalities, she suggested, shared her belief that Giuseppe Balsamo was a something of a rough-hewn saint. After all, an early life of wickedness was not untypical of figures whom the church had now canonized. Giuseppe Balsamo had been no churchgoer, but he did have "a powerful soul." He'd believed that Islam and Judaism were just as holy as Christianity, and that his sect of Egyptian Freemasonry—a form of secular religion cast in the shape of a secret fraternity—could bring all three religions together.

On my last night before leaving Palermo, I met Nina and Sebastian at a restaurant near the market. A few hundred yards along Palermo's main road hung a handsome overhead sign that said *Ristorante Le Conte de Cagliostro*. It showed Count Cagliostro's face in his most famous pose, eyes elevated to the heavens in spiritual ecstasy. Inside, the menu offered dishes in honor of the now legendary characters that Cagliostro encountered. Queen Marie-Antoinette was a Gorgonzola and Casanova, a cream-filled pastry.

The patron looked like pictures of the younger, not yet portly Giuseppe, brown and square with brawny forearms. He explained that the restaurant's name was already in place when he bought the business some years ago. He'd come from the building trade in Naples and knew nothing about Cagliostro. Over a drink one day he'd asked a blacksmith friend what he should do about the restaurant's name. The man said with great vehemence that he must keep it: Cagliostro had been an *uomo di popolo*, a man of the people. What better name could you want?

When locals come to eat in my restaurant, the patron said with a grin, they always make the same joke. They say that the street name of Balsamo's house was changed because everyone in the Albergheria always called it *via Pisciata* rather than *via Perciata*. In the market, when the call of nature comes, he explained, Balsamo's lane is the best spot to go. That's why it stinks of piss. Piss, they say, was Cagliostro's first elixir. They love him for that.

After the meal, I decided to make one last pilgrimage to via Conte di Cagliostro, treading on the day's sad detritus of squashed tomatoes and slimy fish-scales. The hole in the wall was lost in the dark, and the alley stank of piss, as it should. Beyond the alley, high overhead, I could just make out the gray shape of Mount Pellegrino, squatting over Palermo like some ancient pagan god. It was the same view that Giuseppe Balsamo would have had on a night like this two and half centuries ago when he launched his career as a traveling adventurer.

An hour or so before midnight in the year 1763, two men, one middle-aged and prosperous-looking, the other not yet twenty, walked quietly through the outskirts of Palermo, capital of the ancient island kingdom of Sicily ruled by the empire of Naples.

They spoke softly, holding a dim lantern, anxious not to attract attention to their identities or purpose. Fortunately, the heavy layer of garbage on the streets made it possible to move without noise. Shopkeepers joked that nobles encouraged this carpet of dust, dung, and garbage because it gave their carriages a soft ride. But it was street people, not people in carriages, that the two had to watch out for—the numerous beggars, thieves, cutthroats, and prostitutes who infested the crowded old Moorish quarters of the city. If opportunists were to notice that they were carrying a pick and spade, trouble would follow.

After several miles they reached Mount Pellegrino, its two-thousand-foot shoulders hunched over the gulf of Palermo like a sphinx guarding secrets. It was one of these secrets they hoped to uncover. How often over the centuries had Palermo's rulers frantically buried their wealth in one of the mountain's many caves as the sails of conquerors appeared on the sea below? Phoenicians, Greeks, Romans, Moors, Normans, Spaniards: all had landed on Palermo at some time in the previous dozen centuries. Most had stayed to enjoy the subtropical climate, the rich soil, and the deep blue harbor linking Europe, Africa, and Asia. This city of 200,000 people was a living testimony of a hybrid past.

It was likely that Moorish Saracens had buried the treasure hoard that the two men were after. The richest, most tolerant, and most sophisticated of Palermo's many invaders, these Moors had stayed in Sicily for two centuries and stamped their imprint on landscape and people. Doubtless one of their princes had buried the treasure just before escaping back to Africa: perhaps Ayub Temim had stashed it just as the Norman forces of Roger Guiscard breached the nine-gated fortress of Al-Qasr and poured into Palermo on 5 January 1072.

The younger of the treasure-seekers, Giuseppe Balsamo, claimed to know exactly where the treasure lay. An ancient document had revealed the spot, and a dream or vision had shown the contents—a buried cache containing a hoard so rich that its description made the second man's head spin.

Balsamo had told Vincenzo Marano, one of Palermo's wealthiest silversmiths, that the cache included a carved golden cockerel with ruby eyes and diamond-flecked feathers. The value of this piece alone staggered his mind, even though Marano was in the habit of dreaming about hidden treasure. Not that he was alone in this; everyone in Sicily had stories to tell of buried riches. Finding and securing them was the problem. Young Balsamo had told him that the golden cockerel was perched on a pile of golden coins as high as a dung heap. But in his dream he'd also had a frightening glimpse of the demons, jinn, or evil ones who invariably guarded such hordes. The Saracens charged these brothers of the shadow with protecting holy Islamic riches from falling into greedy infidel hands.

This was where Balsamo would prove so crucial. Of course, people had warned him to stay clear of the boy because he had such a bad reputation. Giuseppe Balsamo was undoubtedly a *picciotto*, a street thug and a strutting show-off. He shared the usual background of poverty. His father, Pietro Balsamo, had been a jeweler in Palermo, not too dissimilar to Marano himself, except that Pietro died bankrupt only a few months after the boy was born to Felice Bracconieri on 2 June 1743. Giuseppe was raised with his older sister in a cramped two-room apartment in one of the poorest lanes in the poorest quarter in Palermo.

He'd become a typical product of the *quartiere dell'Albergheria*. This warren of twisting alleys, decaying tenements, and slimy cobblestones housed Palermo's underworld of prostitutes, hucksters, thieves, and cutthroats. The *sbiri*, or constabulary, tried to avoid entering the quarter's dark canyons under the unfriendly eyes of Arab, Turkish, and Jewish immigrants. Young Balsamo led a predatory street gang who terrorized people from other quarters and would on occasion fight with the police. He was said to be so quick with a knife that he'd killed a churchman, but the police had to let him go for lack of proof. All this, Vincenzo Marano knew, but he also knew that the skills the boy had picked up on these stinking streets would prove invaluable for protecting their treasure from thieves and bandits.

Moreover, Marano considered that Balsamo was a cut above the usual Palermo street thug. His penniless mother had received some financial help from the boy's maternal grandfather, then from his well-to-do uncles, Antonio and Matteo Bracconieri. They tried to give the boy a good education. He was schooled initially by a private master, then, at the age of ten, enrolled for several years at the seminary of San Rocco for orphan children. This was followed by two to three years' training as a novice monk at the monastery of the Fatebenefratelli healing order in the inland village of Caltagirone. After that, his uncles paid for further private tuition from an art master in Palermo.

It was said that the boy possessed exceptional intelligence and imagination: everyone had heard of his precocious quickness. He had a special aptitude for chemistry and could also draw with remarkable accuracy. This facility in copying images on paper extended to reproducing handwriting, printing, and insignia. For example, he'd been able to draw for Marano a plausible copy of the Saracen treasure map taken from an original parchment. True, young Balsamo had often misused his talents. He'd gotten into some trouble for forging theater tickets and monastic leave passes. It was even rumored that he'd turned his hand to copying legal documents.

When these creative qualities were allied to a refractory disposition, they produced a dangerous customer; there was no doubt about that. Balsamo was said to have been expelled from every school and seminary he attended, showing complete indifference to whipping or incarceration. His effrontery was legendary. The story was told that he'd managed to persuade the Fatebenefratelli friars to release him from the monastery through a brilliant feat of satirical wit. Despite his record of poor behavior, the friars had refused to let the fifteen-year-old novice go because of his gifts as an apprentice apothecary working in their healing clinic. Eventually, having been ordered to chant the martyrology during breakfast as penance for his many sins, Giuseppe appalled his fellow monks and succeeded in getting himself expelled by substituting the names of notorious local whores for those of the holy saints.

As a former novice Balsamo was thus well versed in the Christian rituals essential to combating treasure demons. Equally important, he'd acquired a considerable reputation as a sorcerer able to manipulate occult forces. Everyone knew that Christian and diabolical spirits competed with each other; a successful treasure hunter had to understand both white and black magic. The genies of the underworld figured as an active and animate presence in the everyday life of every Palermitan. The landscape rustled with them.

The Albergheria markets where Balsamo lived sold as many varieties of magical talisman as vegetables. Folk seers, wise women, herbalists, fortunetellers, astrologers, and amulet sellers all possessed ways of connecting with the spirit world. Magic amulets were the commonest. They helped to ward off the evil eye (*malocchio*), to preserve health, and to bring luck in love. Giuseppe had learned how to inscribe mysterious magical signs on scraps of parchment to be hung on a cord around the neck. He'd prepared specially powerful amulets for both Marano and himself to protect them in the ordeal with the treasure jinns that lay ahead.

Of course, young Giuseppe's reputation as a sorcerer ran deeper than the making of amulets. During the years at the Fatebenefratelli monastery, the boy evidently became something of an adept of hermetic lore. This was the mosaic of ancient Egyptian and Greek magical theory derived from fabled Hermes Trismegistus, a composite of the Egyptian god Thoth and the Greek god Hermes. Under Sicily's two centuries of Moorish rule, Catholic orders like the Fatebenefratelli had become copyists, keepers, and transmitters of Arab manuscripts and recipes conveying the secret knowledge of Oriental magic and alchemy.

Working with Father Albert in the monastery laboratories, the young chemical apprentice Balsamo had learned how to identify and fix salts, and how to use mercury and sulfur (mercury being feminine and sulfur masculine) to underpin the alchemical processes of transmuting base metals. For spiritual and medical prognostications, the *cabbala* was the standard tool, but it was not easy to understand or use. It comprised a body of Jewish mystical doctrine, as well as a system of prediction known as *gematria*. One could arrive at oracular predictions through calculations based on the correspondences between numbers and Hebrew biblical letters or words. Balsamo had produced several cabalistic predictions that seemed auspicious for the success of their treasure hunt. Naturally one also had to take into consideration the influences of the heavens, based on the matrix of powers exercised by seven spiritual genies or angels who ruled each of the seven planets.

Along with this astrological knowledge Balsamo had also learned from Father Albert the formulas and rituals of adjuration for conjuring up good spirits and exorcising evil ones. After his return from Caltagirone, two such feats in particular had gained the boy great local celebrity. On one occasion he'd used a piece of string dipped in holy oil procured from a country priest to exorcise his sister of a demon. More challengingly, he'd one day offered to show a group of his friends what a girlfriend was doing at that exact moment on the other side of Palermo. Drawing a magic circle in the sand and uttering some strange phrases, he somehow produced her image playing a card game, *tresetto*, with two men. Some said that the boy had merely drawn a lifelike picture in the sand; others swore that he had created a hazy spectral vision in the air. Either way, the gang dashed to the girl's house, where the prediction was confirmed in every detail.

Like it or not, Marano knew that Balsamo was the only man in Palermo with the skill to neutralize the demons long enough to grab the treasure. Balsamo had confessed to Marano that he intended at first to keep his dream entirely to himself but eventually realized that he could never wrest the treasure from its terrible keepers without help. Naturally, he'd thought of Vincenzo because of his reputation as a dedicated and fearless treasure hunter.

Balsamo admitted that the risks were considerable. It was known that Saracen treasure guardians could do terrible things. The troops of Azazel or Eblis, the Arabic ruler of demons, were notoriously vicious. They included bicorns and *chichevaches*, creatures grown bloated and loathsome from feasting on human flesh, as well as jinn, elemental Arab demons that could assume deceptive human forms and suddenly tear you to pieces or derange your senses.

Fasting was necessary; he insisted on weeks without meat. As well, Balsamo purchased and sacrificed nine live cockerels (a magical number, as everyone knew). They had to be black, white, and red; three of each color in an expression of natural sympathy with the sparkling jewels of the hoard. He'd also insisted they undertake ritual ablutions of the skin, followed by the application of special herbs, balms, unguents, and holy oils. All this he accompanied with chants in a high-pitched voice, like the Arabs when they prayed to Allah.

Marano had winced when he heard how much money Balsamo needed to purchase the propitiatory magical materials—sixty silver pieces, a small fortune. True, it was nothing compared with the worth of that golden cockerel. He'd handed the money over reluctantly, vowing to keep a close watch on the young sorcerer. But he had to admit that the boy proved exemplary: Marano could not fault the care with which he'd organized the arduous preliminary rituals.

On the night of the hunt, Balsamo had come from his house in the Albergheria dressed in a black clerical cassock that would help him to evoke the good spirits and give protection against the bad. Pellegrino, where they were headed, was a holy mountain, the site of the most sacred shrine of Santa Rosalia, guardian of the city and healer of the plague. They had prayed for her help in any demonic struggle. People from Palermo and its surroundings made pilgrimages several times a year to visit her statue, so lifelike that it seemed to breathe. Santa Rosalia, protect us and help our quest.

Marano found Balsamo an impressive sight in the smoky light of the juniper torches. With powerful shoulders, deep chest, and quick, light-footed movements, here was a boy who could look after himself in a struggle. He exuded strength and confidence, carrying a phial of consecrated oil and sprinkling holy and blessed water from a basin. Balsamo would manage the magic; Marano would do the digging.

As they reached the stony flanks of the mountain, Balsamo stopped suddenly, gestured for silence, and looked carefully around. Satisfied, he indicated to Marano to take a mouthful of holy water as a further antidote to demons. Methodically, he rubbed holy oil onto the silversmith's forehead and gave him something to cover his ears. He drew a circle in the ground — a ring of magic like an invisible cage protected by the good spirits. He began to sing, moving from familiar hymns to the high-pitched ullulating chants of the Arabs. Suddenly he pointed at a spot and gestured to Marano to begin digging. The silversmith swung the pick, working with intense, excited concentration.

Out of nowhere came a crackle of blue phosphoric flame, a sulfurous smell, the pounding of feet, unearthly shrieks: for an instant Marano glimpsed black faces, hairy skins, devilish horns. Blows beat down on his head and shoulders, driving him to the ground, where he lay hardly daring to move. After what seemed an eternity, he took the silence to mean that the demons had retreated to their lair. When he eventually climbed to his feet, there was no sign of Balsamo. As he hobbled, bruised and bloodied, back to Palermo, Marano hoped that the boy had not been abducted and eaten by the demons.

He couldn't even remember how many jinn had attacked them: was it four or was it a dozen?

The next day, when he had recovered somewhat from the beating, he walked to via Perciata to consult his young seer. All night he'd worried what might have befallen Balsamo at the hands of the demons. But a new horror presented itself on his arrival: Balsamo had left town—supposedly to visit his uncle in Messina—and he'd taken Marano's money with him. Humiliated and enraged, Marano swore on his Sicilian blood to have vengeance. The Balsamo boy would never return to this city, not unless he wanted to rot in prison forever. Vincenzo Marano would not forget.

· I ·

Freemason

Freemason: A member of the fraternity called, more fully, Free and Accepted Masons.

Count Giovanni Giacomo Casanova was bored out of his mind. It was March 1769; on doctor's orders the legendary seducer had been trapped for nearly three weeks now inside the Three Dolphins Inn, Aix-en-Provence, recovering from a serious bout of pneumonia.

The visit had begun well enough. He'd promised himself the treat of a spring stay in warm southern France in compensation for having spent an uncomfortable six weeks locked in a jail cell in Barcelona for passing false bills of exchange. After these rigors, he'd anticipated with delight the thought of spending a week or two in the ancient French provincial capital of Aix, near to an old friend, the marquis d'Argens, a philosopher and voluptuary with a similar appetite for women, food, and fine conversation. Casanova was also tantalized by the prospect of attending the city's famous Easter festival, said to be the most saturnalian in France.

The slight discomfort of being given a room next-door to a nosy cardinal was offset by the advantage of the inn's location — close to Aix's elegant main street, the cours de Mirabeau, with its canopy of shady trees and bustling outdoor cafés. A group of young men interested in sensual pleasures made for affable company at the breakfast table; they listened to Casanova's stories as if to holy writ.

Like him, these were adventurers—men who lived on chance and imposture—but they were only small fry; trainees and amateurs gathered in the presence of a master. They hung on Casanova's words because he'd excelled in Europe's riskiest profession. Every substantial town or city swarmed with members of his fraternity. There were no fixed credentials in this business, no certain pitches, no reliable sources of income. The perils were great, the pitfalls many. Adventurers had to live by their wits, charm, and courage, always ready to make a bold advance or a quick retreat. You might find yourself traveling in a magnificent six-horse coach on one day; chained to the oar of a prison galley the next.

At dining tables, gaming houses, salons, and operas Casanova brushed up against scores of half-familiar figures whose newly minted titles were as extravagant as his own—his preferred title chevalier de Seingalt was an utter invention based on letters picked at random from the alphabet. A good adventurer also needed an attractive professional repertoire and remarkable versatility so that he could offer himself as an entertainer or musician, a painter or an actor, an animal trainer or an expert in fireworks. Many, Casanova included, made claims to special knowledge, touting themselves as astrologers, philosophers, inventors, magicians, or healers. A reliable standby was to present yourself as a foreign soldier, a traveling clergyman, or a government official on diplomatic business. Casanova had worked them all.

Despite occasional setbacks, Casanova could think of no freer or more satisfying trade. With most of Europe at peace, borders were delightfully porous; passports were relatively easy to obtain; and a real or forged letter of introduction to some local dignitary would open doors, provided you were equipped with stylish clothes and a good coach. Europe's myriad petty rulers craved novelty and spectacle: all a good adventurer needed was to select the right hook for local conditions. Casanova specialized in sexual charm, magical imposture, and cardsharping, but when necessary he'd turned his hand to organizing lotteries, inspecting mines, and posing as a healer. Since every European ruler, large or small, aspired to imitate France's former Sun King, Louis XIV, and all nobles, however provincial, modeled their dress and manners on those of France, one had only to acquire the language, fashion, and taste of Paris. No one could excel Casanova in dancing the minuet or turning a polished compliment.

At Aix, there was plenty of opportunity to test these skills. Soon after arriving in the city he was invited to dine with d'Argens, who was staying at his brother's château in the countryside. The fare proved splendid. He filled his mouth with Madame d'Argens's exquisite *crostata*, a huge, crusted meat pie, crammed with little sausages, sweetbreads, mushrooms, artichoke hearts, fatted goose livers, and much else. Holding up his end of the bargain, Casanova shone. He set everyone laughing by frightening a Jesuit who dared to criticize his bawdy conversation. The miserable priest blanched with fear when Casanova declared that the pope, currently under election, was sure to be the hatchet man Cardinal Gagnelli, a Franciscan eager to crush the meddling Society of Jesus. Such anticlerical badinage delighted a young Berliner called Gotzkowski, who quietly promised to introduce Casanova to a livelier set of libertines gathering in town every day for "parties of pleasure."

The Easter carnival also exceeded Casanova's best hopes: its masquerades, dances, and buffooneries were the most frenzied he'd seen, and he'd seen plenty. Venice's *carnevale* had been one of his formative experiences. Extending for the last week before Lent, it drew revelers from all over the country. Aix proved even better. On 25 March he cheered the procession through the city center, as the Devil, Death, and Deadly Sins, in comic dress, struggled to avoid kneeling to the Creator. Putting on a mask of his own—no doubt his favorite *commedia dell'arte* disguise, hook-nosed Pulcinella—he prowled through the streets in search of giggling masked beauties. A whirl of "assemblies, balls, suppers, and very pretty

Casanova, age fifty,
by Alessandro de Longhi

girls" made him stay into Lent: one could not bet-ter the intoxicating mix of sex and religion.

But as so often these days, now that he'd reached the age of forty-four, retribution followed on the heels of his keenest pleas-ures. As he was riding back into town in an open chaise after a visit to D'Argens, the icy north wind chilled him to the bone. Instead of going to bed, he then tried to fulfill a boastful promise to deflower a wily fourteen-year-old peasant girl. Locals had wagered that she could position her body so as to defeat the efforts of all comers. Gotzkowski had already failed on an earlier occasion, and Casanova promised to show him how to suc-ceed. But two hours of roaring and working "like seasoned warriors" resulted only in exhaustion. The wager lost, Casa-nova felt abject.

This defeat put him in a black depression. With Gotzkowski watching, it was an uncomfortably public humiliation. Weakened and troubled, he'd suc-cumbed to a chill. Pneumonia followed — it was so severe that, after a week of spitting huge dollops of blood, he took the last sacraments. In the end the insurance proved unnecessary. Still, the mighty Casanova had to be nursed in bed like a baby for eighteen days before the doctor declared him out of danger, and even then he'd been ordered weeks of convalescence at this tedious little inn.

So here he was, lusting to return to his normal life, tired of playing cards, and irritated with the company of his empty-headed young admirers. One morning at breakfast, however, they'd piqued his interest with a tidbit of gossip — some new guests. An Italian pilgrim couple had checked in yester-day evening and had gone straight to bed, apparently exhausted.

Pilgrims, Casanova thought. In this day and age. What an absurdity. Pilgrimage was a thing of the past. They had to be "religious fanatics or rogues." Either way, though, they'd provide some badly needed entertainment.

Young Giuseppe Balsamo had witnessed scores of Easter carnivals: his mother and uncles made sure of that. In Palermo, the festivities took the form of a *ballu di li diavoli*, or devils' dance. In a ritual contest between good and evil, men of the town, dressed as red devils, danced a symbolic struggle with the black-clad followers of the Virgin Mary, Mother of Sorrows. Of course, the black-cassocked representatives of the church had to win the contest; any other result would be unthinkable. In Giuseppe's own life, however, the outcome of the same duel was less certain: sometimes the Virgin dominated, sometimes the devils.

When in 1764 Giuseppe had been forced to flee from that greedy silversmith Marano, he and his two accomplices, a valet and a priest, headed straight to Messina on the far side of the island. They couldn't stay long—the arm of the law would soon have reached them—but they could cadge a few days of hospitality from Giuseppe's well-to-do uncle, Joseph Cagliostro.

Messina was a staging port for boats traveling into the eastern Mediterranean and over to the coast of Africa, so the miscreants quickly sought a passage before the news from Palermo arrived. While waiting for a boat, Giuseppe says, he met a mysterious old magician called Althotas who became his guiding star and teacher for the next few years. More likely, he was romanticizing a continued relationship with his fellow swindler, Father Atansio. The older man had skills to impart, and the novice rogue some things to learn.

It's not certain, either, where the renegades went at first: for a year or so the trail peters out. Some say they visited the island of Rhodes posing as experts at turning hemp into silk—an old scam, this, but new to Giuseppe. Balsamo himself claimed they caught a ship over to Cairo and Alexandria and wandered around those fabulous Muslim cities soaking up the rich culture of the Orient. And why doubt it? Egypt was a regular destination for Sicilian traders, not to mention an exotic target for would-be adventurers — Casanova had undertaken a similar voyage to Corfu and Constantinople when he was twenty-one.

After some months of traveling and living on his wits, Giuseppe eventually headed to Malta in 1765–1766. Here he disembarked on the island's jagged coast and made his way inland to the small medieval town of Medina, where he managed to land a job working as a kind of servant, or *donat*, with the Knights Hospitallers of Saint John. This powerful clerical order, generally known as the Knights of Malta, had been founded in the eleventh century to look after the health of Christian pilgrims heading to Jerusalem. As well as administering medicine, the knights had taken up the sword against the forces of Islam on land and sea, struggling with Moors and Turks in the Mediterranean for several centuries before establishing themselves permanently on the island of Malta in 1530.

With the onset of more peaceable times, the knights had grown into the wealthiest charitable order in the world. Their embassies, priories, and pilgrim inns were dotted over large parts of the European mainland, especially in Austria, Italy, and Spain. On Malta itself the knights built citadels, churches, hospitals, and palaces, especially in the fortified towns of Medina and Valletta. At the time of Giuseppe's arrival, the grand master of the order, Dom Manoel Pinto de Fonseca, moved between the two cities. A haughty Portuguese nobleman, he was known for his love of ceremony, his lavish lifestyle, and his fascination with alchemy.

What brought Giuseppe there? Family tradition, for one thing: his mother and uncles often boasted of a famous Balsamo ancestor who'd been a grand prior of the knights in 1618. Why shouldn't another bright and ambitious Balsamo make a mark with this wealthy order? Giuseppe had the qualifications, after all. Hadn't he spent several years as a novice dispensing drugs for a similar healing order at Caltagirone? He didn't have to mention being thrown out of the Fatebenefratelli for the little joke about Palermo whores.

And he was perfectly willing to exchange his old black cassock for a smarter new one with a red Amalfi cross, as worn by the serving brethren of the knights.

Giuseppe's skills as an apothecary stood him in good stead. Before long he was given a room next-door to Dom Manoel's grand alchemical laboratory. For the next several years no whisper of scandal touched him: the Virgin was once more in the ascendant. A genuine passion for concocting medicines and experimenting to find the elusive philosopher's stone checked his habitual inclination to restlessness. And when, toward the end of 1767, he could no longer resist the urge to return to the Italian mainland, he carried with him warm letters of recommendation from senior officials of the knights.

After staying a few days at Naples, Giuseppe passed rapidly through Campagna to arrive in Rome early in 1768. He immediately presented his credentials to the Count de Brettville, Maltese ambassador at the Holy See, and, through him, to several eminent Roman churchman, including cardinals York and Orsini. The latter, as it happens, was looking for a young man with literary and artistic talent to help with occasional secretarial work. Giuseppe had fallen on his feet again: he could hardly have hoped for a loftier patron. But it was boring work and Rome's temptations were legion. Soon he began to lead a double life, appearing "sometimes in an ecclesiastical, and sometimes in a secular habit." When out of his cassock, he boosted his earnings by selling a variety of shady goods to tourists as they passed through the piazza di Santa Maria la Rotonda on their way to visit the Pantheon. His homemade goods included "Egyptian" love potions and engravings reworked to look like original paintings. None of this was desperately illegal, but he also began to hang out with Sicilian expatriates who were well-known to the police. And it was at the house of one of these, an old friend from Naples, that Giuseppe met a real-life virgin with whom he would dance for the remainder of his life.

She was a ravishing fourteen-year-old called Lorenza Feliciani. Accustomed to the dark complexions of Sicily, Giuseppe had never seen anything

The young Seraphina

like her. She had shining blue eyes, a creamy complexion, and a mane of blond hair. A lissome figure, full red mouth, and teasing manner completed the enchantment. As for Lorenza, it was difficult for an illiterate artisan's daughter from the poor quarter of the Trastevere to resist the advances of this immensely self-confident twenty-five-year-old, with his impressive air of learning and his bewitching stories of foreign travel. He wasn't bad-looking, either, in an Arabian sort of way: coffee skin, strong white teeth, wine-black eyes, and a mass of dark curly hair. Muscular, too: when he moved, his neck and shoulders rippled and his step was quick and light. At the same time, a high forehead, delicate hands and feet, and a thrillingly sonorous voice gave him a priestlike aura. And Balsamo was undoubtedly a gentleman: he accompanied his seduction with an offer of marriage.

Lorenza's father, Giuseppe Feliciani, a devout Roman brass-worker, thought his daughter a trifle young for marriage; but her suitor was obviously besotted, and he did boast the most distinguished clerical connections. As was their habit, Signor Feliciani and his wife, Pasqua, decided to put their trust in the blessed Madonna del Carmine, whom they entreated to guide their daughter through any future difficulties. Dowries were assembled and the pair married at the Felicianis' neighborhood church, Santa Maria de Monticelli, on 20 April 1768. The marriage was witnessed by two of Balsamo's Sicilian friends. The newly married couple initially shared the modest Feliciani quarters along with Lorenza's brother, Francesco. The house, which doubled as a workshop, was situated in the vicolo delle Cripte, adjacent to the strada dei Pellegrini, an area much frequented by poor pilgrims visiting Rome.

The arrangement soon palled. For Giuseppe, it was all too similar to his stifling upbringing in the Albergheria. He quickly grew irritated at the pious stupidity of the Felicianis, while they found their son-in-law too clever by half. Notwithstanding his clerical past, he enjoyed goading them with bawdy and impious talk. Nor did they like the way he was influencing their daughter. He persuaded Lorenza to adopt her second name, Seraphina, and he encouraged her to flaunt her beauty in unseemly ways. Eventually, a serious quarrel forced the young couple to move out altogether, putting immediate pressure on their meager resources.

In trawling for extra income, Giuseppe befriended an out-and-out criminal called Octavio Nicastro, a hard, hook-nosed Sicilian thug wanted all over the place for theft, fraud, and forgery. Nicastro, in turn, introduced him to a flashy nobleman who called himself the marquis Agliata, minister, colonel, and plenipotentiary to the court of Prussia. Giuseppe was delighted when the charming marquis praised both his artistic talent and his clerical connections—so much so that Agliata agreed to make Balsamo his personal secretary and teach him how to turn his drafting abilities to more profitable effect.

Under Agliata's instruction, Giuseppe learned to forge bankers' letters of credit, merchants' bills of exchange, and even military brevets. The last was a particular revelation: in an instant he could become an army captain in the same way that Agliata had become a colonel. The marquis even invited Giuseppe and Lorenza to join him and his associates on a business trip to Germany where Agliata had considerable influence. On the way, Giuseppe could learn another crucial forger's art—how to put his products into circulation without attracting notice. Naturally, there was a price for such generosity: Agliata hinted that he was mightily taken by Balsamo's young wife.

A dilemma, this, but Giuseppe didn't brood on it for long. Like any red-blooded Sicilian man of his day, he adored his young wife; that is, he loved her passionately, he ruled her absolutely, and he treated her like the child she was. Though quick-tempered, he was not, however, a jealous man, as long as his property was safe. And goodness knows, here was an opportunity too good to miss. Gently and patiently, he explained to Lorenza that God had given her a gift of beauty which she must use for their joint good, just as he must use his own gifts of drawing and chemistry. And why shouldn't she have some fun? After all, she was a lusty girl—and Agliata was not unattractive. The marquis would certainly be generous; there'd be gifts of money, jewels, and dresses. A mortal sin? Not at all; as long as the encounter remained merely fleshly and as long as she continued to love and obey only her husband. As a sometime clerical novice, he could assure her that it would be a venial sin at worst, and probably not even that.

In May 1768 the party left Rome to travel through Italy. In a sense both the Balsamos were crossing a Rubicon. Lorenza traveled in a handsome coach alone with Agliata, while Giuseppe, Nicastro, and the rest of the gang squashed into a humbler vehicle behind. As they went along, the entourage picked up some earnings here and there before their partnership came to an abrupt end at a village just outside Venice. Nicastro, jealous of Giuseppe's growing influence, betrayed them to the police, and Agliata disappeared with the money, leaving them to explain their connection with some questionable letters of credit. Quick-witted Seraphina managed to hide the incriminating forgeries down the front of her dress, so the couple were eventually released for lack of evidence. Even so, they were now alone and penniless.

What to do? Casting about, Giuseppe remembered the pilgrim lodges of the Knights of Malta. Even though pilgrimages had undoubtedly declined in recent years, these refuges still dotted the traditional pilgrim roads, especially in Spain. As long as you walked on foot dressed in recognizable pilgrim's clothes, the inns would provide you with free accommodation and food, particularly on the most famous route in Europe, which ran through Italy, France, and Spain to the shrine of Saint James of Compostela in Galicia. On the way, they would of course be expected to beg alms from devout citizens at each town, but with Seraphina's help they might also attract contributions from the not-so-pious.

Casanova never forgot his first sight of the girlish pilgrim slumped in a hotel chair holding a small golden crucifix on her lap. He reckoned her age at eighteen (she was three years younger). In a soft Roman lilt, she introduced herself as Seraphina Balsamo, sounding weary when she described her recent ordeal. She and her husband, Giuseppe, had just completed the legendary pilgrimage route from Rome to Santiago de Compostela. A grimy pilgrim cloak, thick staff, and heavy oilskin enhanced her look of vulnerability. Casanova's instinctive bedroom glance took in every detail: a fine aquiline nose, a sensual mouth, and a pleasingly ripe figure. He even noticed a tiny flaw in the otherwise perfect picture—drooping eyelids that "marred the tenderness of her beautiful blue eyes."

Casanova later claimed to have realized instantly that the couple were adventurers, but his two accounts of the meeting show some doubt about what to make of these "strange creatures." Seraphina's soft oval face "breathed nobility," yet he detected a whiff of sexual calculation about the way that she pulled back her sleeves to show him the bedbug bites on her white arm. Casanova could not be certain, but her husband seemed to connive at her flirtation as he sat quietly sewing onto his cloak the lacing of cockleshells that signified completion of the arduous pilgrimage.

Giuseppe Balsamo looked around six years older than his wife (he was actually twenty-six); stocky, brown, and muscular. His manner was bold and impudent, too damn streetwise for a pilgrim. Yet the face was pleasant enough: a slightly snub nose and dark curly hair framed a pair of black eyes that sparkled with intelligence. Unlike the exhausted beauty, he seemed untroubled by the rigors of the road. After starting to speak to Casanova in poor French, he lapsed gratefully into strongly accented Italian that immediately suggested a lack of breeding. The man claimed to come from Naples, but Casanova knew a Sicilian accent when he heard one.

Casanova never properly explained why he hesitated to take up Seraphina's hint, even when she followed it up later on with an unaccompanied visit to his room. After all, he'd been celibate for at least three weeks and she was a woman to stir the blood. The presence of a husband was not usually a deterrent for him: Casanova claimed to fear no one. His friend the count de Ligne called him Hercules — tall, wide-shouldered, and arrogant, skin as dark as mahogany, nose jutting out like a bowsprit, a sword swinging at his side. His ferocious pride had provoked at least a dozen duels, and he'd won them all. Did he sense in Giuseppe a different kind of toughness — the savagery of a street fighter rather than a gentleman duelist? There was an air of menace about the man, though Casanova couldn't have known that only twelve months earlier Balsamo had knifed a waiter who tried to swindle him at the Locanda del Sol, a hotel in Rome.

Lately Casanova had grown a shade wary of seductively beautiful young women like Seraphina. His sexual pride had never really recovered from an incident in London five years before when he'd been tormented to near suicide by the wiles of a similarly innocent-looking beauty called Maria Charpillon. She'd turned out to be a heartless courtesan working an extortion racket with her family. More recently in Spain he'd been caught in another honeytrap when a woman had acted as bait for a gang of thieves. Typically in such cases armed thugs would appear and grab your wallet. Alternatively, the woman's husband, accompanied by witnesses, would rush into the room, having carefully arranged to catch you copulating with his wife. Outraged, he would then demand money or gifts in compensation.

On the other hand Seraphina's "air of virtue" was so convincing that he half wondered if she might be genuinely devout, particularly because impersonating pilgrims seemed such an outdated form of roguery. It would get you some coarse food and a verminous bed at a roadside lodge, but any financial pickings would always be meager these days. Even in Catholic Italy and Spain, where traditional religious practices were more tenacious, only a few superstitious fanatics still marched on foot to visit medieval shrines. Were these Balsamos actually pious naïfs, Casanova wondered, or simply out-of-touch imposters?

Rogues or not, Casanova never doubted that the couple had made the backbreaking walk from Rome to Spanish Galicia and then to Aix — its rigors were imprinted on their bodies and clothes. They told him they'd traveled initially through Italy via Sardinia and Genoa, and then moved up through France to Avignon and Montpellier. They described the next leg of their tramp as having taken them 150 leagues across two mountain ranges and over the harsh, stony landscapes of northern Spain. Seraphina was on the brink of collapse by the time they reached Aix. Even so, they planned to make a special trip to Turin to pray over the relics of Saint Veronica's handkerchief, which was said to carry the genuine imprint of Christ's sweaty face. They also told him that generous citizens had often given them more money than they needed, so they'd made a point of distributing the surplus to the poor of each town as they passed through. Only through poverty, Seraphina told him fervently, could they expect to gain true merit in the eyes of God.

Their act — for it was that — was so artfully presented that it fooled even an experienced traveler like Casanova. They were just beginning their supposed pilgrimage. They'd dressed themselves in rags in Milan and walked, via Loreto, Bergamo, and Antibes, to Aix. It wasn't a huge distance, so Seraphina's exhaustion was mostly feigned. The couple managed to trick the cynical Venetian count both because they were brilliant performers and because their religious roles lay completely outside his framework of understanding. Casanova wore Catholicism more lightly than his velvet waistcoat. He'd grown up in one of the world's most sophisticated cities and studied at one of Italy's best universities. There, he'd acquired a taste for French philosophers like Voltaire who convinced him that religious institutions were conspiracies for controlling the ignorant. Though he'd been tempted a couple of times by the perks of a church career, boredom or sensuality had quickly caused him to drop the idea. He could think of nothing more ludicrous than marching in rags to visit a Spanish shrine.

If Casanova was baffled by the contradictions of the Balsamos, they were in no doubt about him. Once he'd avoided their honey-trap, the couple came clean about some of their mercenary intentions (though they never confessed to the falsity of their pilgrimage). The day after their first meeting, Seraphina knocked unexpectedly on the door of the count's hotel room— this time to tell him that her husband had a genius for black-and-white, or chiaroscuro, drawing and, more particularly, for reproducing the styles of famous painters. Would Casanova like to see some examples? He would indeed. Guiseppe was soon showing the count his traveling art portfolio, which included several fans that he'd embellished with ink motifs to look like engravings, as well as a version of a Rembrandt drawing that Casanova thought "more beautiful than the original."

When he congratulated the young Sicilian on this talent, however, Balsamo burst out: "Everyone tells me that, and everyone is mistaken. With my talent a man starves. Practicing my profession, I work a whole day in Naples and Rome, and I earn only half a *testone; that* is not a living." Balsamo then proceeded to show Casanova some more profitable ways of using his artistry. Having begged the count to write them a letter of introduction, Giuseppe quickly forged a copy so adroitly that Casanova couldn't tell which was his own original.

So, as he'd suspected, the man was a rogue, and one with a useful talent. Casanova knew the rich possibilities of forgery firsthand. He also knew the risks. One had to be very careful with this particular trade. Forgery threatened the system of trust and credit on which much of Europe's commerce depended, so governments treated the crime especially harshly. Several of Casanova's forger friends had already ended their lives on the gallows.

Nevertheless, Casanova kindly collected some alms from his young hotel cronies to help the couple on their way. The pilgrims departed early the next day, tramping resolutely in the direction of Marseilles. Thinking about the encounter later, the count could still feel a soft spot for lovely Seraphina; Giuseppe, on the other hand, was simply "one of those lazy geniuses who prefer a vagabond life to hard work." He reflected smugly that the casinos, brothels, and prisons of Europe were crammed with Balsamo's kind.

Whether the young man ended up in the galleys or on the gallows, Casanova knew that the Sicilian would never amount to anything.

For the next few years, Casanova's somber prediction hovered continually on the edge of fulfilment—all the more because Giuseppe Balsamo's career so often echoed the count's own.

Casanova seemed right, too, about the relative merit of the two Balsamos. Time and again only Seraphina's interventions saved Giuseppe from penury or prison. Though the couple did enter Spain, they never reached the shrine. By the time they arrived at Catalonia even Seraphina's zeal had run dry, and she was happy to throw off her pilgrim habit and put on the clothes of a Roman aristocrat. Still, it was one of her habitual acts of piety that brought their first break. Confession to a priest in Barcelona led to a sustained period of church support for this pair of traveling Italian "aristocrats" who were temporarily down on their luck. After that, a succession of wealthy noblemen hired Giuseppe to work as an artist or chemist in exchange for regular sexual access to his young wife. This arrangement led to a year of steady work in Madrid, which ended when Giuseppe got into a legal dispute with another artist. Moving to Lisbon, Seraphina again struck it lucky. Here, with Giuseppe's active encouragement, she caught the eye of a fabulously wealthy Portuguese merchant called Anselmo de la Cruce. He paid Giuseppe eight pistoles a time for Seraphina's early-morning visits and also heaped gifts on her. Giuseppe invested all these proceeds in Brazilian topazes, which he decided would sell at vast prices in the booming commercial metropolis of London.

Something about London brought out an odd streak of naïveté in the normally streetwise Giuseppe: Casanova had been affected in the same way when he visited there in 1763. Unfamiliarity with the language, manners, and Byzantine legal system of the British made both adventurers unusually vulnerable. In August 1771, the Balsamos managed to rent a modest house in Compton Street amid Soho's lively Italian community, but elsewhere the couple were helpless without a translator. Giuseppe's hopes of selling paintings to the king of England came to nothing. He did receive a handsome painting commission from an emissary of the emperor of Morocco, for which he should have been paid nearly fifty pounds. But the money never came. A legal bill did.

Poverty drove Giuseppe into the arms of an Italian crook, a Marquis Vivona. He helped them to extract a hundred-pound blackmail fee from a pious Quaker merchant whom they trapped in a compromising position with Seraphina. Vivona, however, pocketed most of the money and also took off with their topazes, leaving the Balsamos bankrupt. Giuseppe soon discovered the harshness of England's debtor laws when a succession of creditors hounded him into jail. Thank God for Seraphina—her sad little figure at prayer in the chapel of the Bavarian embassy moved the heart of a Catholic aristocrat, Lord Hales, who kindly paid Giuseppe's debt and hired him to paint a mural at his country house in Canterbury. Giuseppe of course messed up again—he botched the murals and seduced Lord Hales's young daughter. Time to leave. With creditors pressing, the couple hastily caught a boat to Calais on 15 September 1772.

Because of Seraphina, they were never without resources for too long. On the boat, a French parliamentary lawyer, Monsieur Duplessis, fell head-over-heels in love with her. All the way to Paris Duplessis wooed Seraphina inside his coach while poor Giuseppe galloped ignominiously behind on a horse. Still, it was worth it for a while: Duplessis made his conquest and Giuseppe was funded to set up a laboratory where he happily tried out the experiments from a sixteenth-century book he'd acquired. It was Allesio Piemontese's *Secretes admirables*, one of the most comprehensive occult manuals ever written, setting out detailed prescriptions for making paints, inks, medicines, cosmetics, and magical spells.

Seraphina's affair with the lawyer suddenly took a troubling turn. Deep down, Seraphina was fed up with Giuseppe's bad temper and broken promises, and she found herself genuinely attracted to Duplessis. He was still handsome at fifty, and, despite having another mistress, he pampered her by paying for theater tickets, dancing lessons, and new dresses. Smitten himself, he begged her to leave her brutish husband, "as in France women have this liberty." She could procure a divorce, he said, by filing a complaint with the police that Giuseppe was a wife-beater and forger. This she did toward the end of 1772.

Duplessis sadly underestimated his rival; Giuseppe would on no account let Seraphina go. He loved her, needed her, and valued her as property. Early in January 1773, he countersued, accusing the lawyer of stealing magical recipes, giving him poisoned wine, and abducting his wife. Confronted with these hair-raising accusations, Duplessis caved in: he was terrified of being lodged in Bicêtre prison. Weeping copiously, he abandoned Seraphina. Giuseppe had won, but he was still determined to teach Seraphina a lesson. Don't mess with Sicilians. He wrote to the Parisian police demanding Seraphina's incarceration in the Sainte Pelagie convent for disobedient wives. Four months later, when she was sufficiently chastened, he had her released. Remarkably, neither husband nor wife appeared to bear any grudge. Lovingly reconciled, they resumed their travels.

Sometime in 1774 they found their way to Naples, where they called themselves the count and countess Pellegrini. Was it nostalgia for his hometown, now that they were nearby, that inspired Giuseppe to borrow the name of the mountain that overlooked the beautiful gulf of Palermo? Nostalgia certainly surfaced when he happened one day to bump into his maternal uncle Antonio Bracconieri, who was visiting Naples. In spite of Giuseppe's past peccadillos, Uncle Antonio was delighted to see his nephew again and to meet lovely Seraphina. He quickly agreed to use personal influence with a senior government official in Palermo, the prince di Pietra Persia, to ease his nephew's return.

Soon Giuseppe was once more gazing up at the gray-black slopes of Mount Pellegrino. Not for long, though: he'd forgotten the enduring nature of a Sicilian grudge. He'd hardly had time to borrow money from his sister before the duped silversmith Vincenzo Marano struck. What a sweet moment of vengeance for the long-brooding treasure hunter. Marano's lawyer bundled Giuseppe into jail and advised him to prepare for a long stay. Never had he needed Seraphina more, nor did she fail him. She bewitched the prince di Pietra Persia so completely that he knocked Marano's lawyer to the ground and intimidated the silversmith into releasing her husband. Before Marano could change his mind, Giuseppe and Seraphina dashed to Messina, vowing never to return.

In a curious recapitulation of the previous flight from Marano, Giuseppe headed for Malta. Again, this proved to be an inspired decision, for the Knights of Malta still remembered him fondly. A decade on, he'd also acquired greater poise and chemical expertise, and they treated him like a luminary. The former novice Giuseppe Balsamo now presented himself as a skilled alchemist, magician, and empiric: never again would he work as a petty artist or forger. When, after four or five months, he and Seraphina embarked in 1775–1776 for the European mainland, they were accompanied by the chevalier d'Aquino, an eminent diplomat of the knights and brother of the ruler of Sicily.

Needless to say, Giuseppe's valise was freshly stocked with letters of recommendation from the knights—credentials that were invaluable for making a new round of clerical and noble contacts. As he moved in the next few months through southern France and back into Spain and Portugal, Giuseppe also began to experiment with the idea of taking over elements of the popular magical legends that circulated in comic-like chapbooks read by common people. He was intrigued, for example, by stories of a mysterious dark-clothed magician called Frederico Gualdo who, though he lived in Venice during the sixteenth century, was also found in a portrait painted centuries before. How could one explain such a mystery—reincarnation or immortality? An even older popular legend told the story of the Wandering Jew, a shopkeeper called Ahasuerus who'd been condemned to suffer on earth forever because he struck Christ on the way to Calvary. By Giuseppe's day, the Wandering Jew was also said to have become a world traveler and carrier of occult secrets such as the philosopher's stone and the universal elixir of life.

Giuseppe was particularly impressed with the fantastic stories that clung to a real-life contemporary adventurer a little older than himself, called Count Saint-Germain. This flamboyant magician had pioneered the process of making himself the real-life hero of traditional folk legends. He even pretended to have known Jesus personally. If Count Saint-Germain could do it, why not Balsamo? Casanova, too, had once tried something similar by getting a friend to pose as a reincarnated wizard. Far more than Casanova, though, Giuseppe discovered a real talent for this kind of bricolage—he found that he could create plausible new identities by patching together bits and pieces of other people's. After all, Sicilians had been borrowing from their invaders for centuries.

By the time that Giuseppe and Seraphina decided to take a second trip to London in late July 1776, they'd assembled enough money and style to make a splash. In another echo of Casanova, Giuseppe now flaunted a magnificent military uniform. On arrival, he announced himself as Colonel Pellegrini of the Brandenburg army, occult scientist, and he carried a silver cane with a jeweled repeating watch in the handle. This alone was worth a small fortune; a pity he'd left Cadiz without actually paying for it. His new identity was still a little fluid: depending on circumstances he switched between the names Count Pellegrini and Cagliostro, the latter surname borrowed from his uncle in Messina. Giuseppe thought it had a good ring to it— somehow noble yet mysterious. Seraphina didn't care whether she was Pellegrini or Cagliostro, as long as she dressed like a countess in the best French fashions.

Colonel and Countess Pellegrini-Cagliostro took a house at 4 Whitcomb Street, Leicester Square, with enough room to set up an alchemical laboratory. Being in England, though, again clouded Giuseppe's judgment—the magician, Pellegrini, behaved as naively as had the artist, Balsamo. This time, he advertised himself as a magical consultant with a breathtaking range of occult powers, including the ability, using cabalistic formulas, to predict the London lottery. What's more, by luck or clairvoyance, he actually did so, picking winning numbers for his landlady in the lottery drawn on 14 November 1776. When this news flashed around local coffeehouses, it naturally attracted a swarm of crooks eager to snatch the golden goose.

Colonel Joseph and Seraphina Pellegrini found themselves caught in a web of deceit, increasingly helpless against the machinations of a growing band of confidence tricksters and shady lawyers. The gang's aim was to have Giuseppe imprisoned on trumped-up charges in order to steal his lottery-winning formulas. Their barrage of legal claims and impostures forced him at different times into the debtors' sections of both the King's Bench and Newgate jails. Not a pleasant experience. And even Seraphina could find no Sir Galahad to rescue him.

It was only half a dozen years since Count Casanova had predicted the likely fate of the young Sicilian pilgrim Giuseppe Balsamo, and Colonel Pellegrini-Cagliostro seemed intent on proving him right.

In a brief patch of freedom during this miserable visit, Giuseppe decided one day to give himself a needed diversion by joining a Masonic lodge. On 12 April 1776, Joseph Cagliostro, colonel of the Thirteenth Brandenburg Regiment and prince of Trébisond, was admitted as a Freemason of the Esperance Lodge, number 289, London, under the Rules of the Strict Observance Rite. The venue was the King's Head tavern in Gerrard Street, Soho, run by the Irish publican and Freemason Peter O'Reilly. Not a grand address, this: Soho was the closest counterpart in London to the Albergheria, Giuseppe's neighbourhood in Palermo. A scattering of French and Italian artists and entertainers jostled to earn a living alongside Irish dockworkers, West Indian ex-sailors, Jewish rag merchants, and Lascar street vendors. Only a stone's throw away, the rookeries of Saint Giles seethed with beggars, thieves, and prostitutes.

The King's Head was a magnet for artisan emigrés, and one of the few places in London where Giuseppe and Seraphina could make themselves understood without a translator. The Esperance Lodge, as a result, was a socially modest affair: its members included a hairdresser, a ladies' shoe-maker, a pastry maker, a waiter, a musician, and a couple of painters. Most were French or Italian. The grand master was a French upholsterer known as Brother Hardivilliers. Amid the fug of smoke and the malty reek of ale, the uniformed Colonel Cagliostro and his elegant wife glittered like stars. For them, an additional attraction of the Esperance was its status as a "lodge of adoption," a type of lodge that allowed a parallel woman's organ-ization to operate within or alongside the male one. The Cagliostros were to become "Lords and Ladies of Hope." This was rare, even in France where the adoptive idea originated: in Britain such lodges were almost nonexistent.

The Cagliostros were also lucky to have struck a foreign-based brand of Masonry dedicated to speculative and mystical "High Degrees." Though Freemasonry supposedly had its roots in medieval guilds of working stone-masons, contemporary forms of the movement went back only a century or so. Sometime around the 1630s, it had emerged, or reemerged, in Britain as a chain of speculative secret clubs professing ideals of benevolence and fra-ternity, and using sets of secret signs, symbols, and rites based on the biblical allegory of rebuilding the Temple of Solomon. It captivated ecumenically minded men living in a country split by religious and social war, especially because members could feel like equals within their lodges. In 1738 the British movement split into the "Modern" Rite, giving allegiance to the Grand Lodge of England, and the "Ancients," an independent Scottish breakaway group sympathetic to the Stuart pretenders. By the time of Giuseppe's arrival in 1776, however, both streams had become socially respectable; they functioned mainly as convivial and philanthropic clubs for middle- to upper-class males.

Though spread originally by Britons living abroad, European Freemasonry mutated in line with local or national peculiarities. In Germany, France, the Netherlands, and Sweden, especially, Masons introduced further layers of initiatory grades on top of the foundational British degrees of apprentice, journeyman, and master. Many European versions evolved a grandiose body of extra rituals, degrees, and symbols modeled on medieval legends and ideals, and often known, after their supposed Scottish origins, as "Ecossais Rites." In part, these rococo impulses reflected the presence in the European movement of aristocrats and intellectuals keen to enjoy forbidden liberal atmospheres and values. An initial absence of centralized controls also encouraged the European movement to fragment. A significant minority of new rites adopted mystical programs based on alchemical and hermetic mysteries that promised the complete spiritual and physical regeneration of their members.

The Esperance Lodge in Soho was cast in this last mold. It was a humble branch of the powerful Strict Observance Rite founded in 1754 by a Saxon nobleman, Baron Charles von Hundt, whose lodges had then spread through much of Germany, eastern Europe, and southern France. The success of the baron's rite rested in part on his organizational brilliance; he built up a military-like structure of seven regional chapters. At the same time, he caught the first simmering of a new European mood—a nagging dissatisfaction with the spiritual emptiness of the dominant cult of reason, or Enlightenment, spread by French philosophers. The baron found that he had tapped an unexpected emotional pool: his followers yearned to rediscover artistic, erotic, and spiritual pleasures fed by their passion and imagination.

Von Hundt's brand of Masonry drew on the legend of the Templars, a glamorous medieval tradition of knightly chivalry, spiritual benevolence, and occult power linked with the warlike crusading order of the Knights Templar, a onetime French rival of the Knights of Malta. After being brutally suppressed by the pope in 1312 because of their independence and prosperity, the surviving Templars were supposed to have gone underground in Scotland. Here, they were said to have resurfaced as Freemasons, having passed their magical secrets into the keeping of a mysterious Unknown Master who fashioned seven "degrees" linked to knightly titles and sumptuous ceremonies. Three elements of this Templar myth, in particular, attracted Colonel Cagliostro: the mystique of joining a secret vanguard, the belief that there would soon be a Templar revival, and a Palermo slum-boy's delight in exacting vengeance against the church establishment of his day.

Many Strict Observance lodges had also absorbed a further heady element of medieval mythology that circulated widely in Germany during Von Hundt's day. The legend of the Rosicrucians, or the Brotherhood of the Rosy Cross, might have been made for Cagliostro. This intoxicating story had been invented in a series of seventeenth-century literary texts written by a German Lutheran pastor, Johann Valentin Andreae, who hoped to use them to reform the corruptions of the papacy. *Fama fraternitas* and other writings projected the existence of a shadowy occult brotherhood committed to transmitting the secret lore and practicing the good works of a German medieval wanderer, Christian Rosenkreutz. As the story went, Rosenkreutz had been orphaned when young, had been put to work for several years in a monastery, and had then roamed around Africa and the Middle East learning ancient biblical, alchemical, and medical secrets out of which he forged a manifesto for spiritual and moral regeneration. His fraternity had built a secret "House of the Holy Spirit" to teach adepts how to heal the sick and practice both alchemy and Christian piety.

Before Colonel Cagliostro could claim to be either a Templar or a Rosicrucian, however, he had to undergo initiation by the Esperance Lodge. His fellow candidates on the day were a French valet, an aged Italian musician, and Countess Seraphina. For her, the process was painless enough; she recited an oath of secrecy and was presented with a certificate and a sacred garter to wear at night on her shapely thigh. Its embroidered motto urged her to

"Union, Silence, Virtue." Her husband had a much harder time of it. After he'd intoned his oath of absolute secrecy and obedience, several officials, dressed in caps and aprons, blindfolded his eyes, tied a rope around his waist, and hauled him on creaking pulleys to the ceiling. Suddenly the rope gave way and he crashed to the floor.

His complaints of a damaged hand did nothing to mitigate the ceremony's next phase. Colonel Cagliostro watched uneasily while a pistol was loaded with powder and ball. His eyes were again covered. He was handed the pistol and brusquely ordered to comply with the oath of obedience by blowing out his brains all over the tavern. He hesitated; he heard yells — coward, get on with it — and pulled the trigger. There was a detonation; he felt a blow on the side of his head and smelled acrid gunsmoke. By some miracle he was still alive; and as his panic gave way to clarity, he realized that it had been a ruse: the lodge officials had given him an unloaded pistol and simulated the discharge. Colonel Cagliostro was now a Strict Observance Freemason, and not an ordinary one either; he'd been initiated simultaneously into the first three grades of the rite, a considerable honor.

For Giuseppe Balsamo this was an initiation in the deepest sense — the transformative moment of his life. It marked an end and a beginning, the death of an old identity and the start of a new. This ordeal in a humble pub began a process of transfiguration that would turn the coarse Sicilian caterpillar into a gorgeous Arabian butterfly. Of course, it didn't happen immediately: bruising one's backside on the floor of a Soho pub was hardly a complete conversion experience. But something fundamental happened to the Sicilian adventurer that day in the King's Head tavern. Entry into the secret world of Strict Observance Freemasonry at last gave Giuseppe a framework for his remarkable intelligence and ambition. Masonry became the crucible of his genius. The theology, ritual, and organization of monastic Catholicism that he'd soaked up in boyhood could flow into an institution with all the trappings of a secular church. The artist, actor, and choreographer in him thrilled to the Masonic taste for pageant and theater. His pow-

ers of oratory found their forum, and the conviviality of the lodge gave his
personality room to move as well as a community to care for his needs.

He loved Masonry's arcane symbols: the omniscient or "all-seeing eye,"
the death's-head, the two-edged sword, the phoenix rising from the ashes,
and, above all, the *ouroboros*—an ancient Persian emblem of eternal life and
indivisible knowledge represented by a serpent eating its own tail. As for
Rosicrucianism; with a few adjustments, the story of Christian Rosenkreutz
became his own. The parallels seemed uncanny. Could it be that Colonel Cagliostro was Rosenkreutz reincarnated?

While browsing among the barrows in Leicester Square not far from
the King's Head, Giuseppe picked up a shabby manuscript that fired the
second stage of his conversion. It was a treatise on the Egyptian origins of
Freemasonry, written by George Cofton (or Coston). Little or nothing is known
of the origins of this obscure Masonic text. No one has found the original.
Some say Cofton was a former Irish Catholic priest who developed the model
of an Egyptian rite from the Rosicrucian-influenced doctrines of another
continental mystic, Martines de Pasqually, but he could equally have been
a minor Oxford scholar of eastern religion named George Costard, who
produced several works on ancient magic around this time. To Giuseppe, it
mattered not.

With the help of a translator, Dominico Vitellini, he discovered that all
contemporary Masonic rites had become more or less corrupt. The pure and
unadulterated form of Freemasonry had originated in the mists of antiquity
among the pyramid builders of Egypt: its founder was a mysterious Egyptian high priest called the Great Cophte or Copt, and its rite and ideals far
surpassed all later accretions from Scotland or elsewhere. Egyptian Freemasonry aspired to nothing less than the complete physical and moral regeneration of humanity through reunification with the divine spirits. Was
there ever a nobler ideal?

The idea of an Egyptian rite exploded like a rocket in Giuseppe's mind.
The mystique of Egyptian magic was something this boy from the Alber-

gheria already felt in his bones with the passion of a native. Of part-Arab ancestry, steeped in the Moorish-Sicilian culture of Palermo's poor, educated in Egyptian-based occult learning, a sometime traveler to Malta and North Africa—he even looked Arabian. No wonder Giuseppe instantly saw himself as an Egyptian seer. Why not? Egyptian origins carried the mystery and glamour of the most fashionable part of the Orient without much risk of exposure. Egypt was a huge and faraway place: noble bloodlines, romantic stories, and secret skills were his for the conjuring.

What Giuseppe glimpsed was still only a kernel of possibility; he had not even decided between the names Pellegrini and Cagliostro. But the value of his newfound Strict Observance passport became quickly evident. In December 1777, the couple fled England for Calais: they'd lost most of their money to swindlers and Giuseppe still had the stink of King's Bench prison in his nostrils. Within a few months, however, Colonel and Countess Pellegrini were welcomed by the Strict Observance Lodge of Perfect Equality in The Hague as if they were royals. After flourishing his insignia, Giuseppe gave his Dutch brothers and sisters a stirring speech, and, in return, both he and Seraphina were issued with further Masonic certificates. They departed through an "arch of steel" made by Freemasons with raised swords, the highest Templar tribute.

Who said Giuseppe Balsamo would come to nothing?

Toward the end of July 1778, almost a decade since he'd first met the Balsamos, Count Casanova's dismissive words came back to haunt him when he again bumped into the couple. This time the roles were reversed: in Casanova's home city of Venice, Giuseppe was the count, and Casanova the vagabond.

Casanova, now fifty-three, probably picked up the news, as was his habit, from chatting to clumps of Venetian loungers in front of the Ridotto. Have you heard—there's a foreign gent just driven into town in a laquered coach pulled by six black horses, with liveried servants in front and behind? He must be loaded. Apparently he's called Colonel Joseph Pellegrini, Comte Fenix, diplomat in the service of the government of Spain—a swarthy thickset man in a fancy military uniform.

At some point Casanova saw the colonel close enough to recognize him, despite the newfound elegance. There was no mistaking his luscious companion, Countess Seraphina di Pellegrini, even though her hair was coiled in fashionable ringlets and her arms were covered in French lace. The pilgrim business must be doing well—Casanova observed sardonically in a later recollection—no sign now of lice bites or bad feet from the stones of Castille.

Casanova quickly made contact. Naturally, he was keen to know how the Pellegrinis had managed their social elevation—all the more so because fate hadn't been kind to him since their previous meeting. Frankly, he was down on his luck. He'd quite recently returned to Venice and found it a disillusioning experience. Like Balsamo, he'd long been an exile from his hometown. Twenty years earlier, the tribunal of the Inquisition—that rump of clergy and laymen who enforced the pope's moral laws—had banned him for life. They'd put him in the state prison in 1755 for practicing magical swindles, but he'd cut his way out and escaped over the city rooftops. Exile hadn't worried him at first, but gradually age brought on an aching nostalgia. He'd spent much of the last decade writing pleading letters to the Inquisitors (as well as an ingratiating pamphlet). Eventually, in September 1774, the Venetian procurator, Francesco di Morosoni, granted him official permission to return.

Hardly anyone in Venice remembered him; new acquaintances yawned at his stories, and most of his former patrons were dead. His stomach was too saggy to attract lovers, and his attempts at serious writing failed abjectly. Soon, he'd become so poor that even his best satin suit and velvet knickers looked sadly spotted. After two years of this, desperation forced him to offer his services as a spy to the tribunal of the Inquisitors.

They'd agreed to pay him by the item for information relating to religion, morals, security, and commerce. Proud Count Casanova became "An Occasional Confidant," one of dozens of casual informers who received a few sequins for each acceptable "confidence," or scrap of tittle-tattle. Good information, though, proved hard to come by; his stories of libertine women misbehaving at the opera and a nude painting workshop in Fish Alley didn't impress the Inquisitors. They had also insisted he adopt a new name to suit his lowly calling. He was known as "Antonio Pratolini at the Casino of Marco Dandolo in the court after the wooden bridge near the Corte delle Colonne"—Pratolini, part-time snoop and professional sycophant.

The Pellegrinis could not have come at a better time because Pratolini's employers were losing patience with the silliness of his reports. His first task was to discover how the couple had transformed their fortunes. Conversation with Giuseppe quickly brought a crucial discovery—the Sicilian was now working as a magician. A magician: Pratolini might have guessed. So, magical imposture has turned ragged Balsamo into elegant Pellegrini. Antonio Pratolini was delighted at the discovery. He knew this profession intimately; he'd been using a bogus oracle to levitate wallets while Giuseppe was still an infant. It looked as if he'd at last turned up a crime that the tribunal would take seriously.

Magic stood high on the Inquisition's hate list. Now that the volcanic religious struggles of the previous two centuries had settled into an uneasy coexistence, the Inquisitors naturally turned their attention toward reforming their own constituencies. They regarded spirit worship, the commonest form of magic among the uneducated, as especially dangerous because it encouraged the rise of heretical sects. Among more educated people, however, dabbling in the occult usually led to criminal deception or flirtations with irreligion. In short, magic could be used by swindlers to plunder the unwary, and by fanatics to inflame rebellion. The nearby Inquisition courts at Friuli were prosecuting:

Spells for general magic
Spells for divine necromancy
Spells for therapeutic magic
Spells for love magic
Spells against wolves, storms, etc.
Spells against bullets
Spells to acquire wealth
Other spells
Maleficia — witchcraft.

Pratolini knew nothing about Balsamo's love of magic and alchemy: he thought of the fellow as an opportunistic swindler, rather like himself, who had recently learned how to con people by mumbling a few plausible-sounding spells. Balsamo — he speculated — must have stumbled on a copy of Alessio Pietmontese's *Secreti ammorabili* (*Secretes admirables*) and been dazzled to discover there:

all the arcane secrets of nature: remedies against all evils, including those sicknesses that ignorant physicians believe incurable: the grand properties of the extract of Saturn; elixirs distinguished from each other by variations of color; the operations by which metals are transformed into gold by purging their heterogeneous elements; golden drinks to infuse the blood with the essence of the sun; methods of divining the inner thoughts of those unable to communicate in person; a way of conversing with spirits from heaven, hell, and other domains; something that makes one invisible; another that enables one to speak with the dead and to see them; another which gives one the power of persuading without the need for reasons and another which makes people appear sexually attractive without having to be either beautiful, polite or complaisant.

But now Pratolini needed to find out what the man was up to in Venice. He managed to wheedle out of the couple the information that they were carrying a letter of introduction to one of the city's richest nobles, probably Count Zaguri—someone Pratolini happened to know. As a result, when the Pellegrinis subsequently visited Zaguri to present their credentials, they were surprised to find the man already suspicious, almost as if he'd been tipped off. Their overture wasn't entirely wasted, though: Giuseppe managed to sell Zaguri one of his remaining stock of paintings, a beautiful Rembrandt-like work called *Venus and Adonis*.

Quickly, the Inquisition's shabbiest bloodhound should have dashed off a report to his masters. These swindling Pellegrinis had furnished him with exactly the kind of intelligence he needed to restore his shaky credibility and win the permanent position he so badly wanted. Yet his spy *Reports* in the archives of Venice say nothing about the imposter couple. Why? Could it be that he was caught flat-footed when the Pellegrinis suddenly left town? Was it his fault they disappeared so suddenly—perhaps he'd been a shade too eager when pressing them for information? Or had they simply taken off because they saw no further prospects in Venice? Either way, they'd not left empty-handed; Pratolini later discovered that the bogus colonel had managed to swindle "some poor devil of a merchant" out of two thousand ecus by selling him an already well-known recipe for refining hemp to the texture of silk. A year later the merchant went bankrupt.

Poor Pratolini had blundered again. Needless to say, it wouldn't be too smart to file a report now that Giuseppe had escaped with a successful imposture under his belt. Only a fool would present such crusty bosses with fresh evidence of incompetence—and Pratolini wasn't that.

One further mystery remained. Why didn't Pratolini make something of the fact that Balsamo had become a Freemason? Pratolini's Inquisition masters regarded Freemasonry as a more heinous crime even than magic. It had been banned on pain of death by a papal bull as early as 1738, and reaffirmed several times since. Pope Clement XIII and his successors were troubled by its secrecy and alarmed at its success among rich and powerful Catholics. Worse, they suspected that it encouraged Protestant and atheistic ideas, as well as political animosity toward throne and altar. Jesuits, in particular, campaigned against the Masonic movement with implacable zeal, convinced that it was responsible for the banning of their order from a variety of European countries.

Colonel Pellegrini's Masonic status didn't escape Pratolini's attention, because the spy was himself a senior Freemason. He'd been initiated into the Grand Orient Lodge of Lyons in 1750 and, soon after, taken the further degrees of companion and master at a lodge in Paris. This in itself could have been an incentive for denouncing Pellegrini; conservative Grand Orient Masons like Casanova intensely disliked the subversive Strict Observance Rite. Elsewhere in his writings, he lashed out at Templars and Rosicrucians for taking part in "criminal underplots got up for the overthrow of public order."

Still, becoming any kind of Mason was a significant step for an Italian who wanted to live in Venice. Allegations of Freemasonry were among the charges that had led the Venetian Inquisitors to lock Casanova up in 1756, and their views hadn't softened since. Other parts of Italy, particularly those states still ruled by the pope, could be more dangerous still. Much later, Casanova claimed to have warned Pellegrini, in Venice, "not to set his foot in Rome" because of the ruthless papal ban against Freemasonry.

Had Pratolini said nothing to the Inquisitors in 1778 because he couldn't bring himself to break his oath of silence or betray a brother Mason? Some scholars speculate that he'd earlier worked as a secret agent for the Grand Orient Lodge of Paris, and that his restless European travels were really missions of recruitment. Somehow, though, it's hard to imagine him being serious enough to pursue the Temple of Solomon all over Europe; for most of his life the Temple of Venus had been uppermost in his mind.

Could it be that Casanova simply didn't attach much significance to Pellegrini's Freemasonry, regarding it as subsidiary to the practice of magic? His own motives for joining the movement were certainly worlds apart from Balsamo's: Casanova didn't become a Freemason to save souls, regenerate human nature, or overturn church and state. No, for him Freemasonry was a "sublime trifle" and a "pleasing invention," a type of social club where "men of importance" could gather for recreation and business. Membership was especially useful for an ambitious young man "who does not want to find himself...inferior to his equals, and excluded from their pleasures." Those who "intend to travel" would find that Masonic lodges could serve as free hotels, far more numerous and comfortable than the wretched pilgrim lodges. At the same time, Casanova warned would-be travelers to avoid any Masonic lodge that might contain "bad company" or "bad acquaintances."

In reviewing whether to report the Sicilian, Pratolini also had to consider the uncomfortable possibility that Giuseppe might implicate him as a fellow Mason so as to save his own skin. The Venetian Inquisition would be only too happy to net them both. Pratolini was not keen for the Inquisitors to know that this supposed Spanish colonel was actually a "bad acquaintance" of old, whose career as imposter, magician, and Mason bore a close resemblance to his own. Getting the miserable Sicilian locked in the I Piombi prison would bring little satisfaction if it meant having to join him behind bars.

On reflection, it was just as well that Count Pellegrini and his wife had decided to leave Venice in pursuit of fresh victims. If, as Casanova suspected, they were headed for far more dangerous spots, he could leave it to others to mete out a proper vengeance.

Necromancer

Necromancer: One who by charms can converse with the ghosts of the dead.

Frederick von Medem had been dead barely nine months when, at the end of February 1779, Count Alessandro di Cagliostro and his wife Countess Seraphina arrived at Mitau, capital of the small East European duchy of Courland-Zemgale now called Latvia. Frederick's untimely death had devastated the women of his family—his stepmother, cousin, and aunt and most especially his younger sister Elisa von der Recke.

Elisa's paternal family, the von Medems, were joining royalty, but she couldn't have cared less. After her beloved brother's illness and death, what did it matter that her giddy half sister was about to marry Duke Pierre de Biron, ruler of the duchy? Elisa's attitude toward the festivities ranged between tepid and frosty. Everyone knew that Biron's lowborn father had been given the throne in 1737 only because he'd been a lover of the Russian czarina.

Given the ring of powerful enemies that coveted Courland-Zemgale's strategic Baltic seaports and rich mineral deposits, the duke of Biron's independence was purely notional. Wild tribes of Baltic Letts and Kurs had been the original masters of this little windy lowland spanning the area between the Gulf of Finland in the north and eastern Prussia in the south. In the thirteenth century Sweden had imposed a rule of steel that lasted three centuries. At that point the duchy had changed its name from Kurzeme to Courland-Zemgale and become a principality of Poland—itself a weak and vulnerable country. Pinned between the Baltic Sea and the River Dvina, Courland-Zemgale seemed to invite invasion. To the east, north, and south, the duchy was squeezed between the vast predatory empires of Russia, Prussia, and Austria. In fact, so weak was Poland's grip on its principality that Courland-Zemgale was really a personal plaything of the Russian empress, Catherine the Great. She'd reinstated the original duke of Biron from Siberian exile to the throne of Courland-Zemgale in 1764, then allowed his son Pierre to succeed in 1772. And Catherine made no secret of the fact that she could remove her puppet at whim.

Elisa, a straight-backed twenty-five-year-old who wore her hair pulled away from her scalp and piled in layers on top of her head, felt that she'd outgrown the shallow attractions of balls, fetes, and spectacles, even if they lifted the ducal court from its usual atmosphere of provincial dullness and petty spite. Biron's newly restored baroque palace with its three hundred rooms looked impressive, but his wedding guests were dullards, mostly solid German Protestants of admirable piety but narrow horizons. Mitau itself, with a population of 20,000, could hardly claim to be one of Europe's great capitals. The von Medem womenfolk—Elisa, her aunt, her stepmother Agnes, and her cousin Louise—soon tired of the wedding fuss. They were all avid readers and spiritual seekers, still troubled by Frederick's death. What a relief, then, when the menfolk returned home one day with news that a mysterious alchemist had just arrived in town. Though her father was at first irritatingly coy about the new visitor, Elisa insisted on a full briefing.

The visitor was Alessandro di Cagliostro, and the Masons of Mitau had been expecting him. Elisa's father, Count Johann von Medem, and her

uncle Landmarschal Otto von Medem were
Masonic officials of high rank. Many years
earlier, they'd been initiated in Halle, taking
the title knights of the Royal Order of Poland
of the White Eagle and of King Stanislaus.
As students in Strasbourg and Jena the two
brothers had also worked with several learned
alchemists, who gave them a "penchant for
mystical wisdom and chemistry." The von
Medems had since led a succession of mystically-minded lodges in Mitau, culminating
in their present Strict Observance lodge.

Elisa von der Recke, about thirty

News of Cagliostro's geographical progres-
sion reached them well in advance through
the chain of affiliate lodges in the major towns of northern Germany.
They'd heard, for example, that he astonished a senior Rosicrucian in
Nuremberg by showing him the highest secret insignia of the order, a serpent devouring its own tail. In Leipzig, he'd apparently sent a ripple of
concern through the lodge Minerva aux Trois Palmiers by denouncing
a local Masonic leader as a diabolist and predicting the man's imminent
death. And one of their own eminent Masons in Mitau, Chancellor von
Korff, a former major and spy in the Russian army, happened on 25 February 1779 to have been visiting Königsberg when Cagliostro was passing
through. Although impressed by the man's eloquence, von Korff reported
back his anxiety that this traveler might be one of the pope's secret Jesuit
agents on a mission to infiltrate and disrupt Freemasonry. His colleagues in
Mitau shared this concern.

Since the death of Baron von Hundt three years earlier, the Strict Observance Rite of the Masonic movement had been plagued by internal dis-
putes. A series of pretenders had jockeyed for leadership. One of the most
prominent—Dr. Johann Starck, professor of philosophy in the Mitau

gymnasium—had lived in Courland-Zemgale for several years. He'd called himself Frater Archidemides and tried to introduce a new variant of the Templar Rite. After causing local ructions, he eventually shifted to Darmstadt. Most Masons of Mitau, including the von Medems, decided in retrospect that he'd actually been a secret Jesuit dabbling in black magic.

Not surprisingly, then, Marschal von Medem had been cautious when Count Cagliostro arrived in Mitau a few days later. The marshal was expecting a humbler figure, but Cagliostro was dressed in the magnificent uniform of a Spanish colonel, a bright red waistcoat stretched over his prominent stomach. His hair, though thin on top, curled luxuriantly down to his shoulders, and jeweled rings flashed on his hands. With a flourish, Cagliostro presented his insignia as a grand master and Rosicrucian prince of the Strict Observance Rite and announced with great solemnity that he'd been charged with reforming their rite throughout eastern Europe. His orders came from the legendary "unknown superiors," including the Great Copt, immortal founder of Egyptian magic. These superiors were deeply concerned: the original Egyptian mysteries that should inform the movement had become corrupted or lost and many Masons had fallen into the hands of devil worshipers and skeptics.

These were large and important claims, the marshal responded. Would Count Cagliostro prove his credentials by performing some key spiritual operations in front of the senior Masons of the city? Of course, the colonel responded graciously, he'd expected nothing less. The Mitau officials quickly set up crucibles and an athanor, fired up the ovens, and watched closely as Cagliostro proceeded to transmute a quantity of mercury into an ingot of silver. Impressive, certainly. Still, the colonel would agree that a clever charlatan could stage such tricks—not that they were suggesting anything, but they needed to be sure. Would Cagliostro be willing to undertake a different kind of test as well? Perhaps he might hold a séance to contact the heavenly spirits, as was customary among the Strict Observance Masons of Mitau?

Cagliostro did not hesitate for a second. He asked only for a child under six, free from all taint of vice, to serve as medium for the spirits. Yes, the marschal's youngest son should make a perfect *pupille* (for this is what Cagliostro called his male mediums). Having arranged the Masons in a circle, he sat the boy down on the Landmarschal's knee. Gently Cagliostro anointed the *pupille's* head and left palm with oil of wisdom, chanting a psalm in a deep, sonorous voice. This ritual, he explained, initiated the *pupille* as a seer whom the spirits would then recognize. Soon the child had begun to squirm and sweat: "Ah, a good sign; the spirits take pleasure in this boy." Cagliostro fixed his eyes on the child, telling him to concentrate all his thoughts on the anointed hand. Turning to the marschal, Cagliostro whispered that he should now ask the spirits something. Without letting the child hear, the marschal replied: "Ask what his mother and sister are doing at home at this moment." Cagliostro put the question. Silence. After a while, the child suddenly squeaked that he could see his mother, his cousin Elisa, and his sister Louise. "Oh, Louise is holding her hand to her heart in pain. Now she's kissing my brother Charles."

Seeing the marschal shake his head emphatically, Cagliostro terminated the séance. The marschal explained that the boy's vision was unfortunately pure fantasy: his eldest son Charles Tittemunde von Medem worked at an army base some distance from Mitau and was currently away on exercises for several weeks. Cagliostro was untroubled by this bad news, suggesting quietly that they test the vision anyway. So, the group of senior Masons walked over to the Landmarschal's house: sure enough, Charles himself greeted them at the door. An unexpected change of plans had freed him to visit. And, yes, Louise had been so excited that she'd experienced heart palpitations.

When Elisa heard the story from her father, hope flickered in her heart. Here at last could be the answer to her desperate need to speak with the spirit of her brother Frederick. She and the other von Medem womenfolk immediately agitated to meet this Colonel Cagliostro.

They were not disappointed. Cagliostro's magnetic eyes, dignified bearing, and mysterious words captivated them utterly, notwithstanding his strange mixed argot of Italian, French, and Arabic. Elisa felt an instant and powerful rapport with the man, writing in her journal, "He is the most extraordinary person I have ever met."

When Count Cagliostro hinted after a week or two that the Great Copt had granted him special permission to set up a "lodge of adoption," Elisa was overjoyed. A parallel lodge that included woman members, as sometimes practiced in France, was entirely novel in eastern Europe. The bored and idealistic von Medem women were ecstatic. Here was a chance to extend their knowledge of the occult for the public good: Cagliostro promised to "lead them to the highest bliss imaginable on earth."

Their new lodge opened formally on 29 March under the control of an inner circle of women founders: Elisa, her cousin Louise, her aunt Madame von Medem (née de Kayserling), her stepmother Agnes—all led by Cagliostro's charming wife, Countess Seraphina, the presiding grand mistress. The male officers were Elisa's uncle and father; her cousin Charles; their close family friend Count Otto von Howen, hereditary governor of Mitau; Major von Korff, the current chancellor; and three other distinguished friends: a doctor, a notary, and a lawyer. The lawyer, Councillor Sigismund von Schwander, joined the lodge only to protect Elisa's innocence, because he was both jealous and suspicious of the new spirit summoner or theurgist.

To Elisa's annoyance, her personal spiritual agenda had to take second place to her father and uncle's grosser material concerns. Having tried fruitlessly all their lives to discover the legendary philosopher's stone for transmuting base metals into gold, they were overjoyed to have an alchemist of the first rank to guide their laboratory experiments. But this desire also faded into second place when Cagliostro announced in the lodge one day that he knew of a place in Courland-Zemgale where secret chemical recipes and magical instruments were buried. He added casually that the cache also contained a large quantity of gold, but this was of little interest to him. Pressed by the excited von Medem brothers, the theurgist took up a pencil and sketched an unusual forest copse, which they instantly recognized: it was on the Landmarschal's country estate of Wilzen, where the brothers had grown up.

As boys, Otto and his younger brother Johann had been inspired by the local serfs' legends to try a few treasure hunts in the forest of the estate, but quite apart from their fear of Rubezahl, a demon of lonely woodlands, they'd been put off by the unending scrum of birch, aspen, and elder trees fighting for sunlight among shadows and undergrowth. Cagliostro's drawing was uncanny: how did he know the exact copse? It was a full day's coach ride away, and his eyes could never have seen it nor his feet have touched the forest floor. Ah! exclaimed Cagliostro, it's true, I have never visited the forest in the conventional way, but while everyone else was resting one afternoon, spirit guides—under the command of the Great Copt himself—flew me through the ether to see the spot. It only took half an hour.

From that moment, the male von Medems were obsessed with organizing a treasure expedition. Toward the end of April 1779 three coaches eventually set out in procession from Mitau. They first crossed a sequence of bridges that gave the town the appearance of having floated down from the Gulf of Riga to beach, midstream, among jade and white river shallows; then they rumbled over the cobbled streets of old Mitau past the tree-lined boulevard.

They passed through the far gates of the city; suddenly, they were out in the open. The road snaked in front of them as far as the eye can see, over the flat Zemgale plain toward Lithuania, a brown-soil sea of farmland and tundra uninterrupted for eighty miles except for Lilliputian farmhouses, patches of bog, and scattered copses. At Wilzen, Landmarschal Otto von

Medem was already waiting impatiently, along with his wife, daughter, and two sons, scanning the low horizon for sight of the expedition.

One carriage contained the Landmarschal's younger brother, fifty-seven-year-old Count Johann von Medem, a normally severe man but unusually animated on this occasion. Seated next to him was Agnes Elizabeth (née von Brukken), his third and recent wife—he'd buried two others, including Elisa's mother. The couple chatted in German as usual to their longtime mutual friend and distant relative Major von Korff, once a doubter but now one of Cagliostro's keenest admirers.

Behind them trundled another coach, owned by High Burgrave Count Otto von Howen. The town governor's excitement was divided equally between prospects of the treasure ahead and fantasies about the beautiful woman who sat cuddled up beside him. With the encouragement of her husband, Countess Seraphina Cagliostro had already seduced the Burgrave, and he was completely in her thrall. He'd given her an expensive jeweled ring and was eager to ply her with other gifts.

The leading coach carried Count Cagliostro and Elisa von der Recke. A high rounded forehead, long straight nose, and full squarish jaw gave Elisa's face an assertive look, but her large, moist eyes were fixed devotedly on the stocky, older man. Strange emotional currents crackled back and forward between them. Cagliostro's words tumbled out in a coruscating stream— eddies of French, Italian, and Arabic, sometimes sublime, sometime opaque. He leaned toward her, his hands, as usual, briskly orchestrating his speech. He was trying to coax Elisa into agreeing to something, but she wavered, skittish and nervy, unwilling to agree.

He spoke of spiritual rather than material treasure—about the character of good spirits and the ghostly precincts of the newly dead. The ability to commune with celestial beings, he said, was not restricted to a few sages like himself; it was open to all good people. Under Cagliostro's tutelage, mortals could achieve reunion with the Divine Spirit by completing a journey of the soul through a series of temptations and ordeals.

Elisa knew that he was trying to tempt her with an offer to fulfil her most ardent hope—to obtain news from the spirits about the fate of her late, darling brother Frederick, or Fritz. In exchange, Cagliostro wanted Elisa to come with him on a journey to the court of Saint Petersburg. He flattered Elisa that her intelligence and family connections would seize the heart and mind of the Russian empress, Catherine II. Together, Cagliostro and Elisa would persuade Catherine to become grand mistress of a lodge of adoption based on the original Egyptian Rite. The empress would then launch a sweeping mission to purge the corruptions that infected Freemasonry throughout Russia, Poland, and eastern Germany. Once again, Masons would discover how to practice communing with the good spirits—the lost Egyptian art of theurgy.

Some immortals, he suggested, assumed human form so as to help deserving people master the science of spiritual communication. Jesus, "the first and greatest magician who ever lived," had appeared on earth as a simple carpenter after initiation at the pyramids of Egypt. And of course Jesus, like Cagliostro, commanded the seven heavenly "archangels," "genies," or "good spirits"—Anael, Michael, Raphael, Gabriel, Uriel, Zobiachel, and Anachiel.

Nothing could be hidden from the scrutiny of these seven spirits, Cagliostro warned her. They were instructed to spy on Masonic initiates to stop them from falling into the clutches of necromancers—the evil spirit-controllers of the underworld. Similarly, the spirits were ordered by God, the "great architect," to impose moral and spiritual tests on adepts like himself to find out if they were practicing black rather than white magic. Elisa should be thankful that he'd instructed the good spirit Anachiel to watch over and guide her.

As though staging exactly such a test, Cagliostro suddenly asked her for any gossip she'd heard about a certain Monsieur X. She replied ingenuously that she did indeed know something but could not betray her promise to her mother and two friends by revealing it. Despite Cagliostro's growing insistence, she was obdurate: Elisa had given her word. "Serpent, that I have nurtured in my breast," he eventually exploded, "Swear to me that you know no incident from X's life, that has been confided to only three others beside you."

Too shocked by this outburst to reply, Elisa's eyes brimmed with tears as Cagliostro continued with awful ferocity: "Now, hypocrite, answer me. Why have you fallen silent all of a sudden? So you know nothing about X?" Choking back her emotions, Elisa summoned all her dignity:

> Count, your behavior shocks me. I don't know who you are playing this scene for, since you only have me at your side, and, as you say yourself, I am continually observed by your servant genie, Anachiel. Since I have the eye of omniscience on me, that reads my innermost heart, I don't feel I have anything to fear from Anachiel's survey, as long as he truly is a good spirit. And if he is not then he can tell you whatever he likes about me. I give my trust to the Supreme Master, who can keep all demons and necromancers under rein, and I am certain that he can bring all the world's seeming disorders to an end.

As fast as it erupted, the count's anger vanished. He leaned forward, looked deep into her eyes, and pressed her hand gently:

> You're a good soul. Your silence does you honor. I could never have imagined such strength of spirit and intelligence in one so young. You have extricated yourself from this situation better than I could possibly have hoped. I can now explain to you what it was all about. I was charged by my superiors to test you with this difficult question. Had you confessed the information to me, I would have worried that your weakness would cause you to succumb to future temptations and that you would fail at the cliffs of the highest magic.

At this point a clump of forest indicated that they'd entered the Wilzen estate. For a while Cagliostro read a small leather book in silence; then he began to mutter low prayers in a strange language. Breaking suddenly from his trance, he gesticulated through the carriage window to the driver: "There! There!" Almost before the coach stopped, he leaped out and strode toward a clearing, pointing excitedly at a tree-stump, cracked and blackened by lightning.

When the other coaches stopped and the remainder of the treasure seekers finally arrived, panting, he explained that the hoard had been buried under the base of this tree six centuries earlier. This was not, he insisted, the time to begin digging for it; the demonic forces swirling around the stump were too strong. His body shook with the effort of holding them at bay. They needed to hurry to the manor, meet with his young *pupille*, and begin a series of preparatory séances.

Over the next eight days the disciples watched helplessly as Cagliostro struggled to overcome the malign influence of the treasure demons. It was clear, he said, that a necromancer was blocking them from the treasure in order to raise it himself. While the men dug fruitlessly around the stump and the women gave advice, Cagliostro wrestled with the spirit monsters. He prayed; he paced up and down frenziedly; he chanted from his book. Sometimes he retired for long stretches to his chamber, from which they heard unearthly groans. He ate next to nothing and seemed never to sleep.

Two or three times Cagliostro took the Landmarschal's six-year-old son with him into the forest to hunt for parchments that might break this spell. The little boy remained their conduit to the good spirits, without whose help the treasure seekers had no hope. The child's innocence, Cagliostro explained, was a guarantee of good faith, and a spiritual membrane through which angels could reach ordinary mortals. Each morning, Cagliostro held a preparatory séance with the boy. Sitting away from the others behind a small screen—to aid concentration, Cagliostro insisted—the little *pupille* stared into a crystal bowl filled with water until spectral images shimmered beneath the surface.

One morning the child squealed from behind the screen that he could see the earth opening up at the base of a tree stump. He felt himself being tugged along by some uncanny presence that forced him down steep, twisting steps. Cagliostro told him to control his fear by counting each step out loud. The Masons believed they could hear the child's actual footfalls. With mingled terror and excitement in his voice, the *pupille* said that he was now creeping down a long narrow corridor. Suddenly he shrieked. Yes, he could see fat golden ingots, rivers of coins and jewels, weird magical instruments, yellowed manuscripts covered in strange symbols, a container filled with glittering rose powder.

"Now I see seven beautiful people in white clothes. One has a red heart on his chest, the others have red crosses on theirs. Something is written on their foreheads but I can't read it."

The listeners exchanged glances—obviously these people in white were the seven archangels wearing the battle uniform of Templar Knights. Minutes later, Cagliostro closed the séance by asking the *pupille* to embrace his celestial visitors. The child did so and was kissed in return. Fourteen kisses: the awed Masons distinctly heard this angelic spooning.

By the end of the eight-day ordeal, Cagliostro, haggard and exhausted, declared himself content. He'd not told his disciples earlier for fear of frightening them, but more than they realized had been at stake. Had this buried arcana passed into the hands of his necromantic rival, "it would lead to the saddest outcome for the world...and it would be some centuries before our globe would be cleansed." Thanks to the power of Cagliostro's rituals, and especially of a magnetized nail that he'd buried just under the tree stump, a protective ethereal shield had been erected around the treasure. The necromancer—whom he identified as a Masonic pretender called Professor Starck who'd lived in Mitau a few years earlier—could not in the future come within three hundred leagues of the area. The mention of Johann

Starck excited the family, for they remembered the shady crypto-Jesuit and his attempts to dominate their lodge.

Furthermore, Cagliostro promised that when he attained the highest grade of the Egyptian Rite, which would not be long away, he would have the power to banish the remaining demons, bring up the treasure, and hand it over to the von Medem brothers. No, no, he wanted none for himself, though he eventually bowed to the family's insistence that he accept the parchments and rose powder to give to his superiors.

Though Cagliostro's frightening outburst in the carriage had troubled Elisa, she still wrote excitedly in her journal about "this priest of secrets." Such was her intoxication with the magician that friends worried about her mental health, which had been fragile ever since the death of Frederick. They noticed that Elisa's longing to communicate with the spirit of her brother was growing into an obsession. For nearly a year, she'd been visiting graveyards at night and sending up desperate prayers to the angels to allow Fritz to speak to her.

Why was Elisa so tormented? Her own answer, published much later as an addendum to her journal, told part of the story. Her father, Johann, had married no fewer than three times, and his children experienced all the trauma that can accompany such changes. Because the second countess von Medem had died while her daughter was young, Elisa had been raised by her maternal grandmother. Her brother Frederick, however, remained in the household of his father and stepmother. Pretty, dreamy Elisa had happily filled her solitary girlhood by reading love poetry, romantic novels, and mystical Christian tracts, "for, of course, religion was a passion for me." Then her father arranged her marriage at the age of sixteen to a boorish middle-aged farmer, Georg von der Recke. Without warning, young Elisa found herself transplanted to a run-down manor house at Neuenberg, "torn from the bustle of the big world and . . . immersed in the silent loneliness of the country." There, surrounded by grim dark woods and empty stretches of tundra, she'd grown to hate her husband and the coarse life he'd inflicted on her. Addicted to boar hunting and beer, he expected Elisa to dedicate herself to sex, child rearing, and running his household.

Elisa chose instead to cultivate an intense emotional relationship with her equally lonely brother, "whom I loved more than anything else and who was the joy of my soul." During her five miserable years of marriage, brother and sister passionately exchanged ideas about life, death, and art. Elisa soaked herself in books of melancholy graveyard poetry like Edward Young's "Night Thoughts" and in the sentimental spiritualism of a Swiss Protestant pastor, Johann Kaspar Lavater.

Frederick, unstable, moody, and full of self-loathing, would brood alone in the forest for hours—Elisa had found him there one day weeping inconsolably. He felt utterly alienated from his father and stepmother, who accused him of having a "dark soul" and punished him for his inability to concentrate. He confided to Elisa, "My parents and relatives do not understand me; they say I lack flexibility, that I am strange, that I am aggressive yet weak and tearful." He'd been a prime candidate for the cult of adolescent hysteria and suicide that bloomed so dangerously around Goethe's celebrated novel *The Sorrows of Young Werther.*

The private letters between brother and sister spoke the language of love. Elisa nicknamed him Fritz; she was his beloved Lotte. Only to Fritz would she "dare to open the secret folds of my soul." They bombarded each other with endearments: "my darling"; "my beloved"; "guardian angel of my life." They wore each other's lockets and adapted love passages from *Werther.* "To be so dear to your heart as I am exceeds all other pleasures except for the one of loving you," Elisa wrote typically in October 1777. "To be worthy of you, Lotte, is the chief purpose of my life," he'd replied.

Then Frederick went to study at the University of Strasbourg, where, within a year, he died of fever. Grief plunged Elisa into the vortices of mysticism, but nothing brought consolation. Her attempts to contact angels using the methods of trance, dream, and vision recommended by a famous London spiritualist, Emmanuel Swedenborg, couldn't produce a celestial whisper.

There were events surrounding her brother's death, though, that Elisa wasn't telling anyone, even her journal. She'd run away from her husband in 1776, and her only child, Fredericka, age two, died soon after, in the same year as her brother. (Though she never had another baby, her maternal instincts flowed into fostering no fewer than thirteen orphan children.) Whatever guilt and trauma she felt over the loss of her daughter were compounded by a tragic breakdown of her relationship with Fritz in the months before his death.

As her brother had settled in at Strasbourg, he began to break out of his shell and enjoy himself. Angrily, Elisa lectured him on the dangers of dissipation. "Nothing is more base than wasting one's time.... The novelty of your environment takes your thoughts from me." Worse, he found a woman called Lisettchen, with whom—to his sister's utter disgust—he satisfied his "lust." Elisa lashed out in jealousy, accusing him of moral crimes. But then Fritz was suddenly gone, and Elisa had to cope with the torment of knowing that he'd died rejected by his beloved Lotte.

She worried, too, that her jealousy had precipitated his death. Suicide was not mentioned, but the possibility hung unspoken in the air. Fritz had apparently hinted at something of the sort only a few months before he died. Elisa forced herself to write to his tutor in Strasbourg, Johann Blessig, for details of her brother's state of mind at the end. Blessig's replies were none too comforting: he described Frederick as a lost soul, plagued by depression and terrible headaches. Another of Fritz's friends suggested that depression had actually killed him: "The doctors give as the cause of the illness that many small vessels were blocked because of long-suppressed inner chagrin."

No less troubling was the possibility that Fritz had been driven to dabbling in atheism as a result of Elisa's rejection. Blessig also told her that

Frederick had become so deeply infected by irreligion that he'd written a tract called *Virtue, Immortality of the Soul, and the Reliability of the Bible,* which he'd later destroyed to keep from her eyes. There seemed no doubt that Fritz had been questioning the very existence of God. Elisa trembled at the dangers of such profanity. The soul of her brother was in mortal peril.

Elisa didn't of course tell any of this to Count Cagliostro; but once the excitement of the treasure hunt died down, she begged him to bring her into spiritual contact with Fritz. She was afraid—she told the count tearfully—that some evil necromancer might have driven her brother to his death. At first Cagliostro was reluctant, claiming the task to be beyond his powers, but eventually Elisa's incessant pleas wore him down. Around 12 April 1779, he capitulated. Cagliostro would give her a magical dream to accomplish this celestial reunion, but first he must warn her about the awesome nature and function of holy magic. It was imperative that she approach sleep in the right mood: her preparations would be crucial to the success of the process.

That evening in the presence of all the Masons, Cagliostro gravely handed her father a triangular sealed paper not to be opened until the next day, after Elisa had disclosed the content of her dream. The paper, he said, posed a secret question that her dream would answer. As Elisa left for bed, Cagliostro locked his eyes on hers and ordered her to fall asleep while saying her prayers.

To Elisa's horror she couldn't sleep at all. Shaking with excitement, she lay awake all night praying for unconsciousness. In the morning, haggard and distressed, she found the disciples waiting expectantly. Nothing, she reported tearfully; she'd not even slept, let alone had a magic dream. Though clearly disappointed, Cagliostro graciously blamed himself for failing to soothe her sufficiently. That evening he repeated his calming preparations at length—but with no better results. This time, Elisa's frustration turned to horror. Overtired and half crazed with anxiety, she thrashed in her bed, neither asleep nor awake, haunted by images that "stirred cold shudders in me." The next morning Cagliostro was less disposed to be charitable: he'd expected greater self-control from her, and he intimated that perhaps she didn't deserve the privilege of a visitation from the heavens.

Elisa felt miserable: it wasn't her fault; she'd tried her best. She wanted nothing more than to have the dream. That night, before snuffing the candles, she read passages by the spiritualist Swedenborg on angelic visitations, but by this time her mind collapsed altogether under the strain. Instead of sleeping, she succumbed to a type of hallucinatory fever. Sweating, crying, and shaking, she was too weak even to call the maid. Next morning, Cagliostro seemed neither surprised nor angry. He explained to the lodge that Elisa was simply too frail to experience a magical dream without damaging her health. If she persisted in this quest, her entire constitution might dissolve.

Later, Cagliostro visited the chamber where she was lying in bed, sick and exhausted. He spoke softly, caressingly, taking up her hand: "My good child, I know how you've suffered this night." Although he could easily have transported her to the very heights of holy mysticism, the risks were too great. He had a higher responsibility to her, as well as to the parents and friends who loved her. Gently, he asked Elisa to acknowledge that she was partly to blame for the failure. "Anachiel your good angel, who has been observing you continually, gives me a constant account of your thoughts and doings. You've been trying so hard to speak to your brother. Don't you understand that you have an almost exaggerated love for Frederick? Anachiel tells me that it is this insatiable mourning for your brother that has brought you to the precincts of mysticism and has planted the seeds for magical knowledge in your soul.... The good spirits are not in a position to exert their elevating effect because you are not yet pursuing holy magic for its own sake."

Elisa's health returned after a good night's sleep and, to her relief, Cagliostro promised to contact Fritz's spirit through the less dangerous method of a spiritual séance.

The day was set—15 April. The place was her uncle's house. The disciples sat on chairs within a magic circle that Cagliostro marked on the floor with chalk, then traced in the air with the swish of his two-edged sword. He began the séance by ritually endowing his usual *pupille* with Frederick's baptismal names. Following this, he performed the ancient magical system of *gematria* by scribbling on a piece of paper a series of cabalistic signs and numbers based on Frederick's initials, F. V. M. After the usual anointings, incantations, and foot stamping, he uttered the thunderous magical words "Helion, Melion, Tetragrammaton." Then he began quietly to question the little seer. Elisa could barely contain her impatience.

"What do you see?"

"I see the beautiful boy who showed me the treasure."

"Ask the beautiful boy if you can see the brother of your cousin, Elisa."

"Yes, there he is, I see him, I see him, he is wearing a red uniform."

"How does he look, sad or happy?"

"Oh he looks so happy."

"What is he doing now?"

"He is holding his hand on his heart and he is giving me the most tender look."

Cagliostro turned to Elisa: "Your brother is in heaven."

She stifled a scream and burst into sobs of relief.

This emotional catharsis ought to have brought Elisa peace of heart, but it didn't. Something had spoiled the mediated glimpse of her brother. In fact, an incident at the same séance started a nagging doubt about Cagliostro's moral integrity. As well as contacting Fritz's spirit, he'd included a nasty surprise for her new brother-in-law, Duke Pierre de Biron, ruler of Courland-Zemgale.

The previous evening at a formal dinner in the grand palace, the duke had raised a laugh among his guests with a mocking reference to magic. Cagliostro's eyes had glittered; nothing enraged him more than ridicule.

Later, at the séance, everyone was surprised when the duke made a sudden spectral appearance. Deep in the currents of the crystal liquid, the *pupille* saw the duke trudging along miserably, covered in chains. To Elisa's horror, Cagliostro ordered the *pupille* to tug hard on the fetters connected to a collar around Biron's neck. Sounds of choking and gagging issued from behind the screen, while Cagliostro hissed, "His Serene Highness will not now doubt my power." And so it seemed; Elisa learned a day or two later that the duke, who was not present at the séance, had been seized at that same moment with spasms in the larynx. Cagliostro had even predicted what medicines he would use to cure the pain.

Elisa couldn't get this piece of malice out of her mind; it tainted the relief she felt at her brother's evident salvation. How could a great moral teacher act so erratically within a single séance? For some time Elisa and the other Masons had been overlooking their master's occasional blemishes. After all, he had been raised in Egypt, where very different codes of manners prevailed. But his Arabic background could not excuse outright cruelty. When she raised the matter, Cagliostro replied tersely that Biron's punishment had been inflicted in obedience to the tutelary spirits, who refused to allow their master to be ridiculed. Even so, Elisa persisted: surely the angels didn't order Cagliostro to be so moody and irascible, or to bully disciples who made trivial blunders such as walking out of his magic circle.

Sighing wearily, Cagliostro explained, as if talking to a difficult child. Didn't she understand that all his shows of petulance were deliberate? The spirits had commanded him to adopt shifting personalities in order to test the integrity of potential disciples. How many times had he told her that a theurgist must continually test to see if followers were flirting with the temptations of black magic? Breaches of Masonic rules, such as crossing the magic circle, could tip the balance between good and evil astral influences. Such disobedience might also hurt the theurgist. Hadn't she noticed that

Cagliostro had collapsed in a fit after the séance? Her cousin Charles, who'd strolled out of the magic circle, and Cagliostro himself were lucky to be alive. Paralysis or instant death could follow such transgressions.

Magic circles were far from trivial, he insisted: their power was produced by magnetic action when he sliced his two-edged sword through the air. True, even for him, there remained some mysteries about the operations of magnetic force. In 1778 Franz Anton Mesmer had demonstrated its efficacy in Paris by using magnetized rods to rid patients of inner blockages that prevented life-giving ethereal fluids from flowing through their systems. Cagliostro knew that his own magnetized sword also created a type of ethereal barrier that kept evil spirits outside the circle and good spirits in it: "Maybe we will understand more when we have proceeded further in this sublime science."

Elisa was invariably reassured by these conversations, but then doubts would return. Cagliostro's awful outburst in the coach convinced her that she couldn't possibly accompany him to Saint Petersburg. The moral risks were too great. It was not that she doubted his powers, but she was afraid that he might be in the grip of the very temptations he denounced. Could it be that Cagliostro was himself dabbling in necromancy? Who guarded the guardian? She'd heard him "give voice to demons" in moments of irritation, and she'd seen him writhe on the floor, drumming his heels, as the evil spirits clamored to enter his soul. Who knows what might happen if these fiends were to make their entrance while she was alone with him in Russia?

The prospect of an unchaperoned journey with Cagliostro was all the more dangerous because Elisa sensed that she was sexually attracted to him and he to her. Moreover the countess's amorous behavior with Otto von Howen set no high standard. What if, while temporarily under the influence of the evil one, Cagliostro made advances to her? Could she resist those narcotic eyes and that caressing voice? On one occasion, he'd teasingly taken hold of a paper silhouette of her, saying that he would use this image to bewitch her. The thought was disturbing. What was to stop him from ensnaring her body and soul like one of the lecherous succubi that were said to prey on sleeping women?

Further confirmation of her fears came a week or two after the treasure hunt, when Cagliostro actually boasted one day to the circle of adepts that he could devise potions to make a woman lust for a man against her will. The Masons had been appalled to hear such lewd words from the mouth of a theurgist. During this same lecture he made the profane claim that he himself was a half god, a product of earthly copulation between humans and divinities in ages past. And though he'd immediately explained away these obscenities as moral tests for his disciples, it had left a nasty taste. Around the same time, Elisa and other Masons had witnessed the undignified spectacle of Cagliostro swearing at his servant and chasing him with a cudgel. How could such brutality and spiritualism coexist in the same body?

Cagliostro's worst display of malice had struck at the heart of her own family circle. He laid a public curse on Elisa's feisty eighty-two-year-old maternal grandmother, Countess Constanze von Korff, because, with the typical bluntness of the aged, she'd openly called him a charlatan. For this, she must pay with her life; Cagliostro was implacable in his wrath. He predicted that in twelve months' time, on 13 May 1780 to be precise, the old lady would eat her breakfast as usual and then be struck dead. No amount of pleading from Elisa or the other Masons would change Cagliostro's mind.

Elisa had herself to resist barrages of pleading from her father, a fierce Courland-Zemgale patriot who believed that Cagliostro might be able to persuade the czarina Catherine to ease her iron grip on their principality. If the mystagogue Cagliostro became a favorite of the czarina, her father argued, who knows what benefits might flow to Mitau? Perhaps she might even make Cagliostro the ruler of the duchy, and thus bestow greater independence upon them.

Cagliostro exerted a more subtle form of pressure by trying to shame Elisa into joining his mission. In late April 1779, after one séance which had gone rather awry when the *pupille* appeared confused, Cagliostro retired into an inner room, struggled audibly with demons, then reappeared pale and tearful to announce that the divine spirits sensed the presence of a potential Judas. Like Jesus Christ, Cagliostro was in danger of being betrayed by someone to whom he had given his deepest trust. The disciples looked at each other uneasily, wondering who it might be. Cagliostro gently asked them all to pray with him for the one among them who courted his ruin. He begged them, too, to pray to give him strength in the struggles with the Evil One that buffeted his body and soul every day.

Elisa was torn between doubt and admiration. She admitted to her journal that, notwithstanding his blemishes, Cagliostro's teachings in the last weeks before he departed for Russia reached a new level of sublimity. His principles proved as tolerant as they were deep. Egyptian Masonry, he insisted, must be open to all sincere religious people, including Jews and Muslims, as long as they believed in a Supreme Being and the immortality of the soul.

He urged his disciples in Mitau to revere the Bible equally with the divine truths of the Persian Zoroaster in the *Zend-Avesta*, the myths of ancient Greece, and the Icelandic sacred poetry of the *Edda*. Pythagoras was his favorite of the pagan philosophers, and he thought Solomon a fine magician whose profundity was limited by his material preoccupations. Egyptian Masonry uniquely combined the cabalistic magic of the Hebrews, the astronomical magic of Islam, and the sublime Christian doctrine of the Old and New Testaments. The three crucial chapters of the Bible telling the secrets of magic and immortality had unfortunately been lost to mankind during the Flood, though their contents were known to the greatest magicians of ancient times—Jesus, Elijah, and Moses. These radiant beings had ruled the celestial realms over eons of time, presiding over the birth and death of worlds and of human civilizations.

He admitted to equivocal feelings toward Moses, who was both a divine seer and the enemy of the Egyptians. For this reason Cagliostro pointed the Masons toward the lost teachings of Enoch, seventh of the patriarchs, who had in millennia past built an underground temple to save the deepest mysteries of magic from the deluge. Enoch had marked the temple's location with two snow-white marble pillars so that a future earthly prophet could one day recover these mysteries. (Cagliostro modestly refrained from saying who this prophet might be.) Seven—he pointed out incidentally—was a particularly sacred number that informed their lives through the power of the seven planets, the seven figures of the zodiac, and the seven archangels.

Cagliostro represented himself as the foremost disciple of Elijah, who was his personal tutelary spirit. Like Elijah, true Egyptian Masons could eventually aspire to rise to heaven in human form. Hadn't Elijah been swept up to the ether in a chariot borne by angels? In the highest grades of his rite, Masons reached the realm of magic where the soul was informed by a shadowless light and could soar with the spirits across the firmament of space and the millennia of time. This celestial ascension had already happened to the Egyptian warrior hero Alexander the Great, whose disciples Cagliostro had met personally in Alexandria.

And even before their ultimate ascension, his true disciples had the capacity to rid themselves of the taint of original sin and reach toward spiritual perfection and eternal life. But to attain such moral purity, they must be prepared to overcome terrible ordeals that would be strewn in their path. Only seventy-two Masons could achieve the first grade, forty-eight the second, and so on up to the final grade, where a dozen immortals would eventually reign forever as spiritual rulers of the world. Masons must realize, however, that the devil's necromancers were constantly at work to divert them from this true path with temptations—earthly greed and cupidity. As a rule of thumb, they should trust spirits whose names ended in "er," and shun those whose names ended in "el."

Though only at the third level himself, Cagliostro expected at any moment to experience the "death and rebirth" that would signal his promotion up the ranks. Though seeming to die, he would rise like a phoenix from the hot ashes. He even told them in strict confidence that his name was not really Cagliostro; neither was he a Spanish count. Both titles of convenience had been allocated by the Grand Copt, whom Cagliostro had served for centuries under the name of the Venetian magician Frederico Gualdo.

By the time of Cagliostro's departure, his teaching had done much to allay Elisa's doubts. He even refused to take her string of pearls with him to Russia and enlarge them there using one of his famous chemical recipes. He disapproved of using sacred magic for shallow profit. He also resisted Elisa's blandishments to bump her up into the second grade of his rite, where she could chat with the spirits. Like everyone else, she had to graduate through the proper Masonic degrees. There were no shortcuts, he told her severely:

Before Christ took up the burden of a prophet, or as you would call him, a savior, The Tempter took him to the top of the temple and offered to teach him about all the treasures of the world, but even these could in no way influence his pure soul. It was only then that he was able to render the earth happy through miracles. It is thus with you.... It is necessary for you not to be tempted by treasures, then the great architect of the world will look benevolently on your path to the mysteries and enable you at last to become a great worker for the well-being of mankind.

And although Elisa managed to hold firm on her decision to stay in Mitau, she agreed eagerly to travel with her father to Saint Petersburg the instant that Cagliostro sent word of Catherine's support for their great Masonic mission. When, after two of the most portentous months of her life, Elisa watched the count embrace and bless his sad little group of disciples, and climb up into the coach, her mind was still in turmoil. Could this extraordinary man, who'd brought them such sublime moral teachings, and who'd eased the terrible burden of her guilt, really be a necromancer? And yet...

As the coach trundled off for Saint Petersburg, Elisa von der Recke could not help feeling proudly tearful to be one of Count Cagliostro's "infants of heaven."

· 3 ·

Shaman

Shamanism: The primitive religion of the Ural-Altaic peoples of Siberia, in which all the good and evil of life are thought to be brought about by spirits which can be influenced only by shamans.

Shaman: A priest or priest-doctor among various northern tribes of Asia. Hence applied by extension to similar personages in other parts.

In the second week of June 1779 Count and Countess Cagliostro arrived at Saint Petersburg, capital of the Russian empire. The last leg of the journey between the Baltic port of Riga and the primitive roadside post station of Caporya had taken them through an increasingly depressing landscape of flat marshy plains, so—despite Cagliostro's sadness at leaving Courland-Zemgale—he was relieved to arrive among Saint Petersburg's famous golden domes, baroque palaces, and tall town houses.

This "Palmyra of the north" had been built on one of the most desolate sites in Europe, a marshy, mosquito-infested estuary of sandbanks, mudflats, and islands at the delta of the Neva River. Here, off the eastern shore of the Gulf of Finland, the river froze through winter, only to break its banks in torrential spring floods. But Czar Peter the Great had foreseen a future "paradise." He'd wrested the land from Sweden in 1702, sent an army of peasants to reclaim its oozy soil, and transformed the handful of wooden fishermen's huts into a grand capital. Proclaimed in 1712, it had grown into a model city of over 200,000 people. Talented architects, builders, craftsmen, gardeners, and artists from all over Europe had flocked to turn Peter's dream into an exhibition of glittering urban elegance. By the end of the eighteenth century, the west-facing harbor held a thousand ships—merchantmen from Europe, galleys from the east, and frigates from Russia's growing war fleets.

The Cagliostros' coach paraded down the straight maple-fringed boulevards, past newly built theaters, handsome squares, wide canals, and rococo mansions of wood and brick, toward the heart of Russian imperial power. Within hours, the count and Seraphina rented spacious premises from an Englishman, Lieutenant General Miller, at an excellent address on the *quai du palais*, close to the granite river embankments that housed the palaces of courtiers, foreign emissaries, and generals.

From the end of the *quai du palais*, Cagliostro could see the citadel he'd come to storm—Czarina Catherine II's sprawling Winter Palace, built by her predecessor but newly refashioned by the Scots architect Charles Cameron along fashionable Italianate lines. In defiance of the cold, damp climate, it flaunted Dutch pavilions, heated gardens, and aviaries of exotic birds. Cagliostro saw himself strolling through the three adjacent buildings of the Hermitage, connected by covered ways, where Catherine entertained her intimates; surveying the scholarly libraries, rare paintings, and elaborate scientific displays; and perhaps even himself creeping into the czarina's green-carpeted boudoir, where so many before him had enjoyed Catherine's legendary appetite.

Opposite the palace, however, stood one of Catherine's buildings that he hoped to avoid—the massive Peter and Paul fortress where she entertained her enemies. Over the portal leading through its thirty-foot-high walls, a crowned eagle reminded passersby that Russian monarchs could also rival the west when it came to building Bastilles.

Still, Count Cagliostro wasn't one to waste time contemplating failure. Later the same day, he drove to the house of Major Charles Henry Heyking, a native of Courland-Zemgale, now serving in the Russian Imperial Guards. On being shown inside, Cagliostro triumphantly handed the major a letter of introduction from someone he mysteriously called Brother II, better known as Count Otto von Howen—Seraphina's recent conquest. As an initiate of the Grand Landes Lodge of Berlin, Heyking had no particular sympathy with members of the Strict Observance Rite, but a warm recommendation from an old family friend in Mitau should have brought a friendly response.

Cagliostro was surprised then to find himself received with cold formality. When addressed in broken French, the major replied in impeccable Italian, but with a patronizing air, as if speaking to an ignorant vulgarian. Count von Howen's letter also spoke of Cagliostro's succulent wife, so Heyking agreed to accompany the visitor back to the *quai du palais* to meet her. The major was sorely disappointed. He'd expected to see a gorgeous Italian princess, but in the flesh Seraphina seemed to him worn-out and middle-aged, with red-rimmed eyes and all the grace of a cabaret rope-dancer.

Cagliostro introduced the major as an "illustrious brother...of our order," and then wasted no time gravely announcing his mission: "I have come to see Catherine the Great," he declared, "to spread the great light in this Orient." Specifically, the count intended to reform and advance the cause of true Freemasonry: "I am chief of the princes of the Rosy Cross," he explained loftily, presenting his Rosicrucian insignia to confirm the claim.

Heyking was unimpressed, responding rudely that Cagliostro's Rosicrucian insignia looked like a makeshift adaptation of the commonplace Polish star of Saint Stanislaus. When the count explained that he'd borrowed this insignia from von Howen because his Rosicrucian original had been stolen, the major snickered.

Cagliostro was dismayed. "I forgive you your incredulity and ignorance...because you are only a child in the order, in spite of all your Masonic grades," he said with dignity. "If I wished, I could make you tremble."

"Yes, if you gave me a fever," the major jeered.

"Indeed," huffed Cagliostro. "What is a fever for the Count Cagliostro who can command the spirits?"

In the middle of this frosty exchange, a servant of the Spanish chargé d'affaires, M. de Normandez, was ushered in carrying a written demand for Cagliostro to present his official credentials at the Saint Petersburg legation immediately. The count exploded; this was too much. Who did this petty official think he was? As far as Cagliostro was concerned, he could go to hell.

In Russia this would not be a wise response, Heyking warned; Cagliostro would quickly find himself in front of the czarina's redoubtable police. Taking a deep breath to regain his fraying temper, Cagliostro politely asked the cocksure major to take midday dinner with them. Once at the table they began to discuss the science of chemistry, a subject on which Heyking pontificated like an expert. Losing patience, Cagliostro decided to put the know-it-all in his place.

"Chemistry," he declaimed, "is child's play for those who know alchemy, and alchemy is nothing for a man who commands the spirits. As for me, I have gold (tapping on the ducats in his pocket). I have diamonds (showing a ring of ugly black diamonds, badly mounted). But I eschew all that and rest all my happiness in the empire that I exercise over those superior spiritual beings who preside over man. They are the souls of mortals disengaged from their bodies, whom I evoke to reappear and to respond to my questions."

Seeing Heyking's sarcastic smile, the count continued, "I'm not worried by your skepticism; you're not the first strong spirit I've subdued and pulverized.... In time you'll come to recognize Count Cagliostro and his power," he added quietly.

The following day, the Spanish chargé d'affaires, Normandez, further humiliated Cagliostro by treating him as an adventurer. The diplomat stated brusquely that foreign Masons were unwelcome at the Russian court and among the nobility of Saint Petersburg. He also ordered Cagliostro to desist from representing himself as a colonel in the service of the Spanish crown,

accusing the count of having formerly used the
name Pellegrini, under which he'd absconded
from Cadiz with a valuable silver cane belonging to a Spanish nobleman.

For the newly arrived seer the signs
were not propitious. A stay at the fortress
of Peter and Paul now seemed rather more
likely than an invitation to Catherine's
imperial palace.

Why had Cagliostro been so confident he
could dazzle cold Saint Petersburg and its notoriously ruthless empress, Catherine the Great?

Catherine the Great,
empress of Russia

Casanova had been partly to blame. The Venetian adventurer loved boasting of his extraordinary
successes in Courland-Zemgale, Russia, and Poland during a tour he'd made
in 1765. Traveling as Count de Farussi, without a penny to his name, Casanova had managed to persuade the duke de Biron to pay handsomely for a
business survey of Courland-Zemgale's iron and copper mines. Having
charmed the nobility of Mitau, including Elisa's spiky grandmother, Casanova had then used his letters of introduction from Courland-Zemgale to
befriend some of Saint Petersburg's most influential nobles. These included
the two distinguished Melissino brothers, Pyotr, general of artillery in the
Russian service; and his elder brother, Ivan, court chancellor—as well as
Count Ivan Yelagin, secretary of state; and Count Nikita Panin, president
of the College of Foreign Affairs.

All these nobles were on close terms with Catherine. Though none had
been prepared to take the risk of introducing Casanova formally at court,
they'd invited him to attend a masked ball at the imperial palace. The wily
Venetian immediately saw through the czarina's disguise and followed her

progress around the room. He held back from introducing himself only because of her giant shambling protector, Grigori Orlov, who was said to have strangled Catherine's former husband, Peter III, with his bare hands. A week or two later, Casanova managed to bump into the empress one morning when she was taking her regular stroll in the summer garden on the left bank of the Neva. He'd so charmed her with the brilliance of his conversation that they met on several other occasions in the same gardens.

It's not clear exactly what Casanova had wanted. Count Nikita Panin hinted that Catherine might give him a lucrative post in the imperial service. Very likely, Casanova hoped for service of a different kind. At forty, he was still setting his sights high in matters of love. Catherine, five years younger, seemed striking rather than beautiful, stocky yet gracious, sensual, and deeply intelligent. Power was also a great aphrodisiac, and she had plenty of that. Her sexual voraciousness was legendary, as were stories of her generosity to male favorites, past and present. Virile young guardsmen were said to need large doses of Spanish fly to keep up with her, and a favorite local joke had it that Catherine's canal was the most expensive in all Saint Petersburg. To Casanova's surprise, though, she'd somehow managed to resist him: he had to content himself with the pleasures of a young serf girl whom he bought for a few kopecks.

Never one to doubt himself, Cagliostro hoped to succeed where Casanova had failed. Despite a growing girth, the Sicilian magician was confident of his attractiveness. His success at Mitau in bewitching Elisa had boosted an already healthy self-regard.

But for much of 1779 Catherine wasn't in the mood for love. With the onset of menopause she was experiencing violent mood swings. According to the British envoy, Sir James Harris, she complained incessantly of discomfort from obesity, swollen legs, and shortness of breath. Around the time of Cagliostro's arrival her sexual self-confidence also took a battering as a result of the infidelity of her lover, Ivan Rimsky-Korsakov. Catherine caught the vain young hussar coupling in a palace chamber with Countess

Bruce, a lady-in-waiting who usually road-tested her prospective lovers. Having passed the test with flying colors, Rimsky-Korsakov had been invited back by the treacherous Bruce for a further course. Catherine was heartbroken, at least until early the following year when she found an even lovelier guardsman, Count Lanskoi.

Cagliostro had to pin all his hopes of conquest on Masonic rather than sexual magic. He knew how deeply Freemasonry had taken root among Russian nobles; all Casanova's connections in Saint Petersburg had been prominent Freemasons. The movement had swept through Russia's nobility both because it appealed to existing mystical tastes and because of the paucity of other social outlets. A Russian general, Count Melissino, had successfully founded his own "Melissino Rite," with Catherine as patron. Why shouldn't Cagliostro do the same? He'd learned on the Masonic grapevine that there were Strict Observance and Rosicrucian lodges in Saint Petersburg and Moscow. This would be a start. As elsewhere, the absence of "lodges of adoption" catering for noblewomen would also give him a chance to claim leadership, especially in alliance with the empress herself.

Here, Cagliostro miscalculated badly: Catherine hated Freemasonry. She patronized the genteel lodges of the Melissino brothers and Count Yelagin mainly because they were too patrician to close down. She also wanted to keep an eye on them. By 1779 the autocratic side of her personality had overridden her earlier liberal tendencies. She sensed that the organizing and running of Masonic lodges could encourage democratic leanings among members, even when they avoided open politics. At the very least, a Masonic lodge was a seedbed within which political opposition might grow. As an ardent Russian nationalist, she also disliked the foreign links of many Freemasons, particularly among those lodges that Germans had introduced into the Russian armed services. By the 1770s, a good many Russian Masons gave allegiance to Grand Master Duke Ferdinand of Brunswick. Catherine naturally saw this as an affront to her sovereignty.

She reserved special venom for the occult wing of the Masonic movement.

By temperament and education Catherine was rational to her core. An early infatuation with the ideas of French philosophers like Voltaire and Diderot had bred in her a deep suspicion of religious mysticism, a distaste that informed all aspects of her life. She loved the clean geometric lines of Saint Petersburg because the city exuded rationality; Moscow's church-filled landscape made her physically ill. Reared as a Protestant princess in the austere rational atmosphere of German Pomerania, she found the Russian populace hopelessly bogged down in superstition. Chevalier de Corberon, a French emissary at this time, summed up the feeling of many foreigners when he wrote that "in Russia you are simultaneously in the fourteenth and eighteenth centuries. But even the civilized part is only civilized on the surface. It is a nation of clothed savages."

Catherine had converted to the Russian Orthodox religion for political reasons, and she valued the church only as a useful arm of state control. She loathed the atmosphere of marvel that accompanied mystical Masonry: it was both ridiculous and threatening. Spiritual ecstasy and questing for the philosopher's stone were at best infantile, at worst a potential source of cults and disorders. As a German princess who'd overthrown a legitimate czar in 1762—her husband, Peter III—Catherine had been troubled throughout her reign by fanatical pretenders who whipped up unrest in the name of the traditional order. In the time-honored manner of peasant revolts, such pretenders collected disaffected social and ethnic groups by issuing apocalyptic manifestos and heavenly promises.

By far the most serious of these outbreaks had been the Pugachev Rebellion of 1773. Emilien Pugachev, a great black-bearded Yaik cossack, had appeared out of the southern Urals claiming to be Peter III reincarnated, and he'd called on discontented Muslims, cossacks, and serfs to overthrow the "devil's daughter" in a holy mission against all things German. Pugachev's rebellion generated a swath of rapes, atrocities, and deaths in the east and southeast of Russia before the leader was eventually betrayed and sent to Moscow for beheading and quartering.

In 1779, Catherine had not yet come to believe that Freemasonry was explicitly subversive, but she worried that Cagliostro might turn himself into a type of Masonic messiah. She later told her German literary confidant Melchior Grimm that "Cagliostro came at a good moment for him when several Masonic lodges wanted to see spirits." And if spirits were floating around, she particularly didn't want her son Grand Duke Paul to see them. He'd already shown a disturbing interest in mystical Masonry. Catherine believed that Paul was exactly the kind of weak-minded dreamer who would succumb to the hazy doctrines of a wonder worker like Cagliostro.

Worse, she thought her son spiteful enough to use Masonry as a basis for dynastic scheming. Spies were reporting the alarming news that Prussian diplomats and Masons had begun wooing the young grand duke. Suspicion that her son could be plotting a coup was one of the few things that could rattle Catherine's steely nerves. Family-based palace coups were of course an occupational hazard for Russian rulers, and Catherine's own gory past — it was said that she'd murdered both her husband and the boy claimant Ivan VI — stoked her paranoia.

Cagliostro probably didn't realize, either, that he'd already committed an unpardonable crime in Catherine's eyes by dabbling in politics in Courland-Zemgale. A group of nobles in Mitau who disliked the sycophantic policies of the reigning duke, Pierre de Biron, had rather vaguely canvassed people to see whether Cagliostro would be willing to replace the man as ruler. Some of them had perhaps hoped that Cagliostro might sway his fellow Mason Frederick of Prussia into asserting German control over their ethnically and religiously divided duchy. If so, they were dreaming. Courland-Zemgale was virtually a Russian province; no change of ruler could be contemplated without Catherine's consent — Frederick himself didn't dispute that. Thus, the haziest rumor of Cagliostro's flirtation with duchy politics was enough to arouse Catherine's fury. A few patriots in Mitau, like Elisa's father, had hoped that the great Egyptian theurgist would so dazzle the czarina that she'd give him Biron's throne as a reward. They could not have been more wrong. Even Cagliostro had rejected the overture, either because he realized that it was a pipe dream or, as Seraphina claimed, because he was frightened of being discovered.

Discovered he surely was. Catherine kept a close eye on Courland-Zemgale; its long Baltic coast was an essential step in Russia's westward expansion, and the duchy made a useful buffer against the ambitions of Prussia and Austria in the east. When Catherine restored Duke Ernest de Biron to the throne in 1763, she'd forced him to sign an agreement to have no dealings with Russia's enemies. This extended to Freemasonry: a French diplomat reported gossip in March–April 1776 that Frederick had ruffled Catherine's feathers by trying to take control of Masonic lodges in Courland-Zemgale. To stop this, she'd toyed with the idea of giving the throne to her foreign adviser and secret husband Prince Grigori Potemkin. Eventually she changed her mind because she didn't altogether trust the master schemer—better a spineless puppet than a potential rival. Instead, she'd contented herself with having her spies monitor politics in Courland-Zemgale with particular care.

Threats to Catherine's imperial ambitions brought out her most ruthless streak. Potential troublemakers soon found that they'd attracted the attention of the "Secret Expedition," her grim cadre of secret police led by Stepan Sheshkovsky, an artist at wielding the Russian knout, or whip. Cagliostro didn't know it, but he was already a marked man by the time he rumbled into Saint Petersburg. Young Heyking's hostility stemmed in part from rumors already in circulation that this bogus Spanish colonel and Egyptian Freemason was actually a Prussian spy. The speed with which Normandez delivered his ultimatum suggests that he'd also been primed.

Poor Cagliostro—he'd captured Catherine's attention all right, but not quite in the way that he'd hoped.

◆　◆　◆

Cagliostro's confidence didn't dent easily. He found a warmer reception among the same group of Masonic nobles who'd entertained Casanova fifteen years earlier. He was welcomed by General Melissino of the Melissino Rite; by Secretary of State Yelagin, who'd gathered fourteen lodges under his personal control; and by Count Alexander Stroganov, an officer of the Grand Orient Lodge of France.

Most enthusiastic of all was Marc-Daniel Bourrée, chevalier de Corberon, the French chargé d'affaires. This witty, acerbic thirty-year-old former soldier was working hard to counteract Catherine's pro-English sympathies, and he would have been wise to avoid entanglement with the suspect new visitor. But de Corberon's addiction to mystical Masonry transcended his diplomatic ambitions. A longtime member of a Lyons Martinist order known as the Elus Cohens, he'd also joined the Strict Observance Templars in 1777. He was so excited by Cagliostro's presence that Catherine called him, tauntingly, a "determined voyager of the spirits."

Naturally Cagliostro was asked by local Masons to undertake a demonstration of his occult mastery. For many Russian lodges, healing rather than spiritual prediction was the favored test. The Rosicrucian legend had always had a healing dimension, and adherents of the rite took it for granted that an elite Rosicrucian like Cagliostro would possess some such skill. Medicine was not, of course, new to him: as a trained apothecary he'd long produced his own balms, tonics, and aphrodisiacs. At Mitau he made up some health tonics for Elisa, and he talked about his philosophy of medicine at some length. He told her, for example, that in Medina he'd been taught a completely different approach to medicine from that practiced by European doctors. He'd learned to diagnose from not only the pulse "but also the color of the face, the look, the gait, and each movement of the body.... The illnesses themselves...have their seat principally in the blood, and in its distribution; it is to this that the doctor must give most attention." Even so, his main calling was as a summoner of divine spirits or theurgist.

Before long, however, senior Russian Masons presented Cagliostro with a direct and unavoidable request to show his skills as a Rosicrucian healer. As in London and Mitau, he rose immediately to any Masonic challenge. In this case he himself was surprised by the results. De Corberon noted that Cagliostro unhesitatingly cured the senior Mason Count Stroganov of a severe and long-standing nervous disease. Other instant successes followed one after another, triggering in Cagliostro a process of creative self-discovery as a healer. The ability to cure the sick not only added a vital dimension to his Masonic repertoire but also deepened his character and identity. Medicine became the third magical acquisition, alongside alchemy and spiritual clairvoyance, in his transformation from Giuseppe Balsamo to Count Cagliostro.

By the standards of the day, Cagliostro's medical knowledge seemed as much "scientific" as magical. He gave de Corberon a recipe for "a delicious distilled water," and in the 1930s the French medical historian Dr. Lalande (Marc Haven) collected and analyzed nine original prescriptions for a variety of health pills, balms, salves, and elixirs that Cagliostro issued around this time. They included an herb tea purgative, a face pomade, a cough elixir, stomach pills, turpentine oil pills, pills from the balm of Canada, two purgative powders, and saccharine oil. In Lalande's judgment, they were all either innocuous or beneficial.

Unlike quack doctors of his time, Cagliostro rarely made wild claims for the chemical value of his nostrums. From the moment he began treating the ill, he insisted that his real curative powers came from divine assistance. Without this, he said, any pills or nostrums he prescribed were of only minor medical significance. Was he merely hedging his bets, or did he really believe that he had a special channel to the spirits? Sometimes he used no medicine at all, merely incorporating healing into his spiritual séances. In these cases, *pupilles* or *colombes* became the agencies through which the spirits did their curative work. Sometimes he simply commanded illnesses to vanish. Whatever method he chose seemed to work, even with the most serious diseases. De Corberon recorded that Cagliostro healed Senator Ivan Yelagin, Saint Petersburg's most influential Freemason, of an unspecified illness; he brought a Madame Bourtouline to a successful childbirth against all odds; and he cured an assessor named Ivan Islieniew of a cancer when other doctors had given up hope.

Cagliostro's extraordinary self-confidence led him to take on the most difficult medical or psychic challenges and to use the most improvised methods. In an exceptionally rare piece of self-testimony, he told a contemporary biographer that he cured a dangerous madman in Saint Petersburg in the following manner:

One of the ministers of the Queen of the Russians had a brother who had lost his reason and believed himself greater than God. And nobody could resist the violence of his rages, and he cried in a loud voice, uttering threats against the whole world and blaspheming the name of the Lord.... And this minister begged me to heal him. When I came unto him, he was possessed with rage, looking at me ferociously and lashing his arms, for he was confined in chains, he seemed to wish to throw himself on me. And he cried: "Let him be hurled into the deepest pit who thus dares appear in the presence of the great God, of him who commands all the gods and drives them far from him." But I, putting aside all fear, approached with confidence and I said to him: "Dost thou linger, deceitful spirit? Is it that thou dost not know me, who am God above all the gods, who am called Mars, and seest thou this arm wherein there is all the force to act from the highest heavens to the depths of the earth? I come to thee to show thee pity and to do thee good: and so it is that thou greetest me, not thinking I have the power to heal, but also that to reduce things to nothingness." And then I gave him such a blow that he fell backward to the ground. When his keepers had lifted him up and he was a little calmer, I ordered that I should be given a meal and I proceeded to dine, preventing him from eating with me. And seeing that he was humbled, [I] said to him: "They say salvation is in humility, to be deprived of all power before me, come and eat." And after he had eaten a little, we climbed both of us into a carriage and went out of town to the banks of the Neva, where the keepers had on my instructions prepared a boat and were seated on the bank. When we were on board, they began rowing and the boat moved off. Then, desiring to throw him in the river so that the sudden terror would aid his recovery (there were people posted to come to his rescue) I seized him suddenly, but he grasping me in turn with his arms, we both fell into the water, he trying to drag me down to the bottom, and I, being above him, overpowered him with my weight, and after a struggle which was not short, I managed skilfully to disengage myself and I rose swimming from the water; he, pulled out by his keepers, was placed in a sedan chair. And when we had returned and had changed our clothes, he said to me: "In truth, I know that thou art Mars and that there is no power equal to thine and I will submit to thee in all things." I answered him saying: "Neither art thou a rival for the Eternal, nor am I Mars, but I am a man like unto thee. Thou art possessed by the demon of pride, and that drives thee out of thy mind: I, I am come to deliver thee from this spirit of evil, and if thou dost wish to submit to me in all things, thou wilt become as a normal man." And from that day he began to let himself be treated, and so he returned to himself, this man whose reason had lost itself in wild fancies.

The apparently miraculous nature of cures like this soon made Cagliostro a subject of excited gossip in lodges, salons, and newspapers. The *News of Saint Petersburg* also reported the sensational fact that this foreign healer refused to take any money for his services. A wealthy courtier, Prince Golitsyn, and his wife had tried to heap money on Cagliostro after he cured their mortally ill baby by nursing it in his own house for twenty-three hours.

The exhausted healer waved aside their offer, explaining that his action had been dictated by simple human sentiment.

Was the count moved by genuine altruism or simply by a flair for self-publicity? Certainly he possessed the latter in abundance, as he showed in a celebrated confrontation with Catherine's personal surgeon, Dr. John Roggerson. Early in 1780 Roggerson called at the house in the *quai du palais* to challenge the wonder-worker to a duel: Cagliostro had humiliated him by curing one of his patients. The doctor, an irascible Scot who had been trained in Glasgow, would not tolerate such impertinence from a mere quack. Cagliostro again described the encounter:

> I shall tell you of what happened in Saint Petersburg. The physician of the Empress of Russia hated me, because I had demonstrated his ignorance, and he came to my house crying: "Let us go out and come and fight with me." I answered him: "If you come to challenge me as Cagliostro, I call my servants and they will come to throw you out of the window; if you challenge me as a doctor, I shall give you satisfaction as a doctor." Affrighted, he responded: "It is the doctor that I challenge." And indeed I had at my orders a great crowd of servants. Then I said to him: "Well, we do not fight with the sword; let us take up the weapons of doctors. You shall swallow two capsules of arsenic which I shall give you, and I shall swallow the poison which you will give me, whatever it is. Whichever of us dies will be judged by men to be the pig."

Roggerson retreated with his tail between his legs.

Roggerson may have challenged Cagliostro at the personal urging of the empress. She habitually used the surgeon for a variety of dubious special services. These included inspecting her prospective lovers for venereal disease before they were allowed into the royal bed and prescribing cantharides for them afterward if their performance was unsatisfactory. Roggerson was undoubtedly one of Catherine's chief sources of gossip about the activities of Cagliostro. The French diplomat de Corberon also believed that Catherine's doctor was simultaneously working as a spy for the British diplomat Sir James Harris.

After he'd humiliated Catherine's physician, Cagliostro's next action bordered on subversion. Early in 1780 he began treating the poor of Saint Petersburg for free. Catherine was aghast. Like most of the *ancien regime*, she distrusted philanthropic appeals to the common people because of their potential to generate political unrest. By turning to the populace, Cagliostro had inadvertently branded himself a democrat. The visitor's Masonic performances among the nobility had been bad enough, but Catherine was appalled by this new populist turn. Through it, Cagliostro could transmute himself into another Pugachev, using his shamanic status to summon an apocalyptic following. The iron czarina was reaching the limit of her tolerance.

Why did Cagliostro take an action that so fatefully and permanently altered his image and behavior? He had not previously tried to reach out to the class of his birth. Catherine's rebuff evidently goaded him into a gesture that would compel attention and protest against his failure. It's possible, too, that he was genuinely fascinated by the Russian shamanic tradition. Traditional folk sorcerers from the Ural-Altaic regions of Siberia commonly practiced popular healing along with other spiritual conjurations, and they were held in deep regard by the people. The profession combined Cagliostro's three great skills: magic, religion, and medicine. Given his tendency to take on local coloration wherever he traveled, why should he not become a shaman?

The poor of Saint Petersburg may also have touched a deeper vein of nostalgia and compassion in the Sicilian. Not far from the *quai du palais* stood a block of arcade markets much like those of the Ballaro adjacent to via Perciata in Palermo. After a short walk or sleigh ride from his grand house in Saint Petersburg Cagliostro could encounter a deeply familiar world. Kalmucks, Tartars, Turks, and cossacks haggled in guttural accents as they bought or sold lentils, beans, bacon, flour, shabby secondhand goods, old tinwear, tobacco, beer, and cheap wooden icons. Here the planned city of Saint Petersburg allowed its Oriental energy to bubble to the surface. There were Turkish merchants in baggy trousers, Muslim women in veils, bearded peasants in shaggy sheepskins, and Siberian shamans dispensing strange medicines. People like this had helped make Giuseppe Balsamo, and now the famous Cagliostro could give something back.

There is no detailed description of his first popular healing clinic, but it was probably much like the one described by a disciple in Strasbourg some twelve months later:

Imagine…an immense room, filled with unhappy creatures almost all without any means and holding up to the sky failing hands, that they raise with difficulty to implore the charity of the Count. He listens to them one after the other, not forgetting a single word, leaves for a few minutes and reenters, carrying a mass of remedies, which he dispenses to each of the unfortunates, repeating at the same time what they have said to him of their affliction and assuring them that they will soon be healed if they faithfully follow his instructions. However, the remedies alone would be insufficient. It is also necessary to recommend bouillon soup so they can acquire the strength to support themselves. Few of them have the means to obtain this soup, so the stock of the Count is shared among them. It appears inexhaustible. He is happier to give than to receive; his joy is shown by his sensibility. These unhappy souls, pierced by gratitude, with love and respect, throw themselves at his feet, embrace his knees, calling him their savior, their father, their lord.

Catherine gave her own more jaundiced description of Cagliostro's Russian clinic in a play she later wrote, *The Shaman of Siberia*. Her Cagliostro figure is a half-crazed shaman and Freemason called Ambane-Lai, who encourages "a mass of all sorts of people" to visit the free clinic he has set up in a genteel hotel in Saint Petersburg. The mob presses through the gates and crushes around him, soaking up his opaque predictions and clamoring for his bogus prescriptions. Eventually rioting breaks out. The police have to be summoned to control the crowd and to carry off the shaman for interrogation.

Catherine also had a more personal reason for her dislike of the visitor. It had to do with her beloved "golden cockerel," the great one-eyed bear Prince Grigori Potemkin, her sometime lover, perennial adviser, secret husband, faithful diplomat, fearless general, sexual procurer, and emotional prop. "Darling Grisha," as she usually called him, had taken a shine to the Cagliostros—rather surprisingly, because he had not up to now shown much interest in Freemasonry. Though it amused him, he'd never taken the movement seriously enough to join a lodge. But the great sensual "cyclops" loved novelty, and Count Cagliostro certainly provided that. Potemkin encouraged Cagliostro to heal several of his sick friends, including one in a hospital on the other side of the city, whom Cagliostro cured without even leaving the house in the *quai du palais*.

Countess Seraphina provided Potemkin with novelty of a different kind. The frequency of the prince's visits to her house in the *quai du palais* soon became a public scandal—so much so that Catherine was moved to make an acid comment on the subject. Then came the sensational rumor that an anonymous noblewoman had offered Seraphina 30,000 rubles to give up Potemkin and leave Saint Petersburg. Apparently Potemkin laughed uproariously at the news: he was flattered by the size of the bribe, but he urged Seraphina to pocket the money and stay.

Gossips suggested that the bribe had come from Catherine. Modern biographers doubt it: the czarina was no longer involved with Potemkin sexually, and anyway she was not usually jealous of his amours. Maybe so, but Catherine was in an unusually troubled state in 1779–1780: the shock of Rimsky-Korsakov's betrayal had hit her hard. At such moments of crisis, she generally looked to her golden cockerel for consolation. A fragmentary letter from Catherine to Potemkin around this time suggests that they were enjoying a sexual relationship, and that Cagliostro was much on her mind. She joked coyly to "Grisha" about the titillating effect of "Cagliostro's chemical medicine which is so soft, so agreeable, so handy, that it embalms and gives elasticity to the mind and senses—enough, enough, basta, basta, caro amico, I mustn't bore you too much." Still, jealous or not, it was not like Catherine to waste money bribing someone whom she could just as easily have thrown into jail. Could it have been Countess Bruce who, in an effort to redeem herself with the empress, tried to bribe Seraphina to leave?

Catherine had sound reasons of state for disliking Potemkin's intimacy with the Cagliostros—most importantly, the friendship could interfere with her adviser's decisions on foreign policy. If, as rumor had it, the visiting Freemason really was a Prussian spy, he might influence Potemkin to work against the czarina's pro-English policies. She loved her Grisha, but she didn't trust him: the prince was an inveterate intriguer. It was time for Catherine to act.

Catherine's animosity toward Cagliostro became public on 24 January 1780 when the government-sponsored *News of Saint Petersburg* announced the publication of an officially sanctioned satire leveled at the visitor. Cagliostro knew that the work had Catherine's strong backing, but he probably didn't realize that she'd actually written it herself. *The Secret of the Anti-Absurd Society* lampooned Masonry as a game for grown-up infants. Its silly rituals and childish rites were borrowed from the "tales of nannies." Catherine's Anti-Absurd Society, by contrast, was Masonry's antithesis—the kind of enlightened movement that Catherine would gladly have led. Eschewing all mystical paraphernalia and occult beliefs, it directed its appeal only to individuals who committed themselves to the pursuit of reason and common sense.

The place of initiation must be a room that shows no resemblance whatever to either a bar or the stall of a quack on a public square. The novice enters together with a member of the society. They knock at the door. He is asked what he wants and replies that he wants to enter. The gatekeeper opens the window and the master of the lodge asks: "Why do you want to enter?" Reply: "Because common sense is guiding me." Question: "Do you feel strong enough to persevere on the route to straight thinking?" Reply: "Examine me." (The door is opened.) The novice enters with his companion, without blindfold and fully dressed because it is considered impolite and not appropriate to be naked during an honest conversation.

Though generally light in tone, the parody contained one serious accusation — it claimed that Russian Masons were sending money abroad and, worse still, receiving payments from foreigners. Catherine was known to view such activities as bordering on treason. Having read the satire, Cagliostro's friends in Saint Petersburg panicked: they were certain that these charges were leveled specifically at him. They urged him to abandon his plans to proselytize among Moscow's Rosicrucian lodges and to get out of Russia while he still could. Seraphina, who'd exhausted the interest of Potemkin and hated the advancing cold weather, was only too ready to swell their pleas. Cagliostro had received a warm request to visit Poland from some wealthy Masonic nobles; wasn't it time to move on?

Catherine later boasted to a friend that she'd expelled the Cagliostros from Russia in 1780, but the couple actually anticipated her intention. In April 1780, armed with letters of recommendation from General Melissino and Chevalier de Corberon, the Cagliostros set off from Saint Petersburg for Warsaw, thankful to leave behind the lengthening shadow of the fortress of Saint Peter and Saint Paul.

◆　◆　◆

Cagliostro and Seraphina soon wondered why they'd battled so long in the hostile environment of Saint Petersburg when such a welcome was waiting in neighboring Poland. Warsaw seemed a delight—fifteen years earlier, Casanova had thought the same. With around 100,000 people, it was a perfect size, and the ancient quarter of the city exuded a quiet beauty. And even if the cathedral and castle tucked into the high bank of the Vistula were no match for Saint Petersburg's golden domes, the Polish capital boasted a matchless king in Stanislaus II.

What a king! It was true that some of the local Polish magnates disparaged him as a flunky of Russia, a glorified steward's son who'd been given the throne as a gift for his feats as Catherine's lover in 1755. He was said to have capitulated to the czarina in 1773—1775, allowing her and her thuggish fellow monarchs of Prussia and Austria to dismember Poland by taking a third of its land and two-thirds of its population. But what else could Stanislaus have done, with Poland surrounded by powerful and predatory empires? His country had virtually no army, a flat frontierless terrain that invited invasion, and a population fissured by religion, ethnicity, and social caste.

Cagliostro and Seraphina cared nothing about this: they thought it good enough that Stanislaus looked and behaved like an ideal king and welcomed them with open arms. Still handsome at forty-eight, he was said to be the best-educated man in Poland. He'd traveled in France and Britain, knew the languages and literatures of both countries, and was a keen amateur scientist and an immensely generous patron. He collected costly paintings and furniture, and he planned architectural innovations with such flair that people spoke of a Stanislavian style. This distinctive blend of French, Italian, and local neoclassical motifs could be seen at its best in the lavishly remodeled grand palace in Warsaw, the newer Ujazdow castle on the edge of the city, and his sumptuous Lazienki gardens complete with English-style lakes, Moorish minarets, and Chinese pavilions.

This was no tyrant who would imprison you at his whim. In fact, the constitutional position of the Polish king was so weak—nobles could veto any legislative proposal in the local Seym (parliament)—that Stanislaus had little choice but to concentrate on the recreational and social sides of life. Witty and charming, he was the most easygoing of hosts. He showed none of the duke de Biron's stiff pretensions that had so alienated Cagliostro in Mitau. And, as someone who both loved and liked woman, Stanislaus instantly put Seraphina at her ease. Though steeped in French rational thought to the point where some thought him a deist, he was also fascinated by the world of the occult. His allegiance to the Masons encompassed both orthodox and mystical versions of the movement. Three years earlier he'd secretly joined a Strict Observance lodge, and he welcomed Cagliostro's reforming mission on behalf of this rite. He was even disposed to believe Cagliostro a supernatural being.

In such a hospitable environment, the visitors flourished. Seraphina quickly gathered around her a little court of Polish noblemen, including the king. They showered her with gifts of jewelry and competed among themselves to stage an extravagant fete on her birthday.

Cagliostro, too, was at his most impressive: his dignity and eloquence pleased nearly everyone. He immediately began to operate a popular healing clinic, working for almost the only time in his life in comfortable cooperation with local doctors and hospitals. He didn't bluster and rant at doubters, as he'd done with Major Heyking in Saint Petersburg, but wooed them with charm and demonstrations of occult skill.

In one celebrated instance a cheeky young princess told King Stanislaus that the Masonic visitor was a fraud. Cagliostro responded by giving her a gentle lesson in clairvoyance. According to several contemporaries, he "brought her from incredulity to admiration" by showing knowledge of the most intimate details of her life. When she then pleaded for a prediction of her future, he detailed a bizarre encounter in the near future that would lead to love and marriage. Friends later reported that the princess duly married in circumstances identical to Cagliostro's prediction.

Prince Adam Poninski, the magnate who had originally invited the Cagliostros to Warsaw, proved a splendid host, insisting that they stay at his luxurious town house. A founder of the Strict Observance lodge "Charles of the Three Casks," to which the king also belonged, Poninski welcomed Cagliostro as a master of his order. He and his Masonic associates were also eager to learn the lost Egyptian mysteries that Cagliostro espoused, and they pressed him to open a lodge of his rite. Within weeks he was conducting Egyptian Rite séances, teaching chemical experiments from the recipes of the seventeenth-century alchemist Frederico Gualdo of Cologne, performing healings, and giving spiritual lectures to a group of enthusiastic noblemen and -women.

But, like the von Medem brothers in Mitau, these Polish Masons began after a while to press Cagliostro to show them how to make silver and gold. Poninski had a particularly greedy streak. As royal treasurer and former majordomo to the king, he'd made himself immensely rich in serfs and land, but now he wanted money. Throughout his adult life he'd experimented fruitlessly to find the philosopher's stone and transmute base metals into silver and gold. Cagliostro's Masonic teachings and spiritual séances with a buxom sixteen-year old *colombe* were impressive, but they did nothing to satisfy Poninski's alchemical longings.

Soon he issued an invitation to the count and countess—to the latter with winks and leers—that they dared not refuse. He insisted that they stay at his nearby country estate where Cagliostro could work on transmutations under the best possible conditions. Poninski's mansion at Wola, a small village half a mile from Warsaw, was equipped with a fine alchemical laboratory. Poninksi would supply Cagliostro with all the materials he needed, as well as an assistant skilled in alchemical methods. Under no time constraints, supported and watched by loving Masonic disciples, Cagliostro would be able to accomplish even the most complex transmutations, including, of course, the production of gold.

Poninski could hardly wait. Alchemists of Cagliostro's reputed skill were extraordinarily rare. One of Cagliostro's French disciples, in a letter of 1781, described the difficulties of producing gold:

> This operation, which appears so simple, is fogged by a thousand names and a thousand different processes. When they [alchemists] perceive that the matter begins to emit some vapor rising to the top of the container...they call this operation sublimation; when the vapors fall down they call this distillation; when they see the heat melt the matter, when it becomes muddy and black, they call this corruption or putrefaction; and when it begins to whiten they call this incineration. I mention only the principal operations, because there are many others. The matter does not spend long on the fire before it demonstrates a considerable change.... Ripleus says that after having seen an infinity of different colors in the primary matter one sees it become white like the snow, then afterward a beautiful citrine, and finally it becomes the color of the red poppy.

When Cagliostro heard the name of his assigned laboratory assistant, he wasn't pleased. It was Count Mosna-Moczynski, former Polish royal equerry and director of the royal theater, now designer of the king's gardens, keeper of the King's collections, and a senior administrator in Lithuania. This self-important, sarcastic nobleman fancied himself an alchemist and liked to cast doubt on Cagliostro's abilities. Moczynski boasted that he would watch the visitor like a hawk for any sign of chicanery.

Cagliostro showed no outward concern. He'd begin, he said, with a demonstration of the relatively quick and straightforward transmutation of mercury to silver. On 7 June 1780, he assembled the Masons at the manor house to perform the operation under their eyes. In shirtsleeves for ease of working, both he and Moczynski also wore full aprons to protect themselves from acid. Cagliostro had asked Moczynski to prepare in advance a pound of quicksilver and some extract of "Saturn" (lead), and also to distill some rainwater until a few grains of sandy matter remained. These he called "virgin earth" or *secunda materia*. Moczynski described the ensuing process:

> After all these preparations were complete he went into the lodge, and he entrusted me with the task of carrying out the whole experiment with my own hands. I did this under his instructions in the following way: the Virgin Earth was put into a flask, and half the quicksilver spread over it. Then I added thirty drops of the lead extract. When the flask was then shaken a little, the quicksilver seemed to be fixed or amortized. I then poured lead extract into the remaining quicksilver, but the latter remained unaltered. So I had to pour two lots of quicksilver together into a larger flask. After I had shaken the quicksilver for some time, however, all assumed a uniform consistency. Its color turned a dirty gray. The whole was now shaken into a bowl which it half filled. Cagliostro next gave me a small piece of paper, which proved to be only the outer wrapping of two others. The inner one contained a glittering carmine-colored powder, weighing perhaps one-tenth of a grain. The powder was shaken into the bowl, and Cagliostro then swallowed the three wrapping papers. While this was going on I coated the bowl with plaster of paris, which had already been prepared with warm water. Though the bowl was already full, Cagliostro took it out of my hands, added some more plaster of paris, and pressed it firmly with his hands. Then he gave it back to me to dry over a charcoal fire. The bowl was now placed in a bath of ashes over the wind furnace. The fire was lit and the bowl left over it for half an hour. It was then taken out with a pair of pincers and carried into the lodge. The bowl was there broken, and in the bottom lay an ingot of pure silver weighing fourteen and a half ounces.

Cagliostro was besieged by excited Masons, congratulating him on the triumph, while Moczynski looked furious. The Pole was certain they'd been tricked in some way. He considered the evidence. The base matter must have come from somewhere else, because the ashes of the fire had definitely not been hot enough to melt it. Come to think of it, Cagliostro had spent the evening before in Warsaw, supposedly communing with the spirits. Moczynski had also noticed that a large container of carbon and some distilled water had disappeared. That's how he must have done it! Cagliostro had prepared a duplicate plaster of paris mold, known as a "hermetic egg." Aided by the dim light, a dark backcloth, and his huge apron, he'd somehow managed to substitute this duplicate egg containing a silver ingot. But how to prove it?

Under pressure from Poninski, Cagliostro now agreed to undertake the more complex and protracted experiment of producing gold. Rarely, if ever, had this most secret of alchemical processes been witnessed. First another hermetic egg had to be prepared. Cagliostro explained that the egg would change color gradually as it moved through seven further alchemical passages, beginning with the gradual acquisition of a black tint and culminating in a brilliant poppy-red when the gold was ready. He predicted that each passage would take at least six to eight weeks.

Time dragged. Each day Cagliostro inspected the egg and added a few drops of eau de forte (acid). After that, Moczynski said, he usually gave the lodge instructions "on various chemical operations, such as making the quintessence of wine by putting it in manure, or obtaining the essence of gold by dousing it in the wine spirit and burning it down with mercury. He taught us the property of oils, such as that of talc...the secret of fabricating pearls." Later he also taught them how to prepare parchments so as to make a *pantagone,* or pentangle charm, which would fend off evil spirits.

Every day he also traveled by cabriolet into Warsaw to "see his ailing ladies" and run his healing clinic for the poor. Moczynski suspected he was actually preparing another duplicate egg. Maybe so; but there was no doubt that Cagliostro was also effecting cures. With each new success, his reputation as a holy shaman spread further into the Polish provinces and brought a fresh crop of supplicants to seek him out in Warsaw. This, and preaching to his Masonic disciples, was what Cagliostro liked to do most, but he couldn't escape the expectations of the Wola circle.

Bored at the slowness of the process, the Wola Masons passed the time in fun and games with Seraphina, who, Moczynski observed, "did not forget to highlight the power of her own charms." Rising late, they generally spent the afternoons romping with her in the woods, supposedly looking for treasure. Returning at sundown, they would dash into the laboratory, recite some psalms to ward off possible evil spirits, then inspect the egg minutely for evidence of changes in color.

Moczynski used the time to pry and look for evidence of trickery. One day he announced exultantly that he'd overheard Cagliostro and Seraphina discussing how she had aided Cagliostro's substitution during the previous transmutation by tossing the original flask and egg out of the laboratory window into a cesspool. Moczynski's fellow Masons dismissed the story as spite; they'd been watching closely and had seen nothing. But then luck began to turn the critic's way. On the evening of 16 June, the young *colombe* whom Cagliostro had been using for his séances in Wola started shrieking out that he'd made improper advances to her—evidently Cagliostro had been engaging in some fun and games of his own. A few of the Masons were shocked. Most just shrugged their shoulders; they were men of the world. But just a minute; this was more serious: Moczynski, having quizzed the girl, now claimed that the séances had been rigged. Cagliostro had given the *colombe* written instructions in advance and had shown her how to fake the sound of angel kisses using her own forearm.

Even the most devout disciples began to show their concern, but Cagliostro turned on his full histrionic powers. He was willing, if necessary, to perform alchemical experiments with heavy chains on his feet and a seal on the door to prove his probity. He further swore to kill himself with his own dagger if they were not totally persuaded by the time the hermetic egg reached its fourth passage. Just as opinion was swinging back in his favor, Moczynski played his trump card. Scouring through the garden, he'd found remnants of a broken flask from the earlier transmutation; part of it had missed the cesspool. Poor Cagliostro felt that he'd fallen right in the pool himself. Now his best eloquence was wasted: the Masons of Wola saw him as a fake.

Eventually, he began to curse them all. They were unworthy of the name of Egyptian Masons, and their avarice would one day be their undoing. He'd waste no more time on such jeering idiots. On 26 June 1780, he and Seraphina packed their bags and returned to Warsaw.

Even so, it was not easy to get rid of Cagliostro. He still had many supporters in Warsaw, especially among those whom he'd healed, including some very influential noblewomen. He also continued to honor his agreement to provide prescriptions and medical assistance to the hospital of Salerne. Everything turned on whether he could retain the favor of good King Stanislaus.

Whatever his own inclinations, the king was in an awkward position. For one thing, he relied heavily on the political support of great magnates like Poninski and Moczynski. More important, he'd learned during the interlude at Wola that the empress Catherine strongly disapproved of Cagliostro. This decided him: Catherine's whims were law, and he needed no reminder that a large contingent of Russian troops was posted permanently at the Polish border.

And if truth be told, the sentimental Polish king still adored Catherine. He always recalled the way she'd looked when they first met in Saint Petersburg in 1755:

She was at that perfect moment which is generally for women who have beauty, the most beautiful. She had black hair, a dazzlingly white skin, long black eyelashes, a Grecian nose, a mouth that seemed made for kissing.

But while Stanislaus remained forever nostalgic, the empress lived in the world of realpolitik. She was known to boast that she'd put him on the throne because his complete lack of entitlement to the position made him all the more dependent on her.

Stanislaus couldn't be anything but a weak monarch; Catherine used him like a conjuror's dummy. On top of this, Poninski and Moczynski, having returned from Wola, began poisoning other influential Warsaw magnates with claims that Cagliostro was a crook and his wife a whore. They even ostentatiously offered the couple some money to leave, as if they were mere adventurers. Cagliostro rejected the bribe with contempt. But those Poles who admired him as a great healer could do little to help—too many of them came from Warsaw's humbler ranks. Stanislaus had no choice but to deny the couple access to court and to let it be known that they must leave his sad country. Sometime between the end of July and September 1780, the count and countess slipped quietly out of Warsaw.

Catherine's attack on Cagliostro turned out to be prophetic, in a way. She'd smashed his Masonic mission to Russia and helped hound him out of Poland, but in the process she forced him to discover a potent new talent. He'd been no shaman when he arrived in Saint Petersburg, but he left as one.

Count Cagliostro also left Catherine's vast imperial domains an angry and embittered man. As he and Seraphina headed through Germany in the autumn of 1780 en route to try their luck in the famous mystical mecca of Strasbourg, they passed through the city of Frankfurt. Here, Cagliostro's Masonic insignia attracted the interest of two officials from the topmost echelons of the Strict Observance movement. They were also, Cagliostro later claimed, secret members of the Bavarian Illuminati. This was a gen-

uinely conspiratorial secret society. It had been founded in 1776 by Adam
Weishaupt, a brilliant former Jesuit law professor from the University of
Ingolstadt, who hated the church that had shaped him. He'd founded a
tough disciplined cadre of republicans dedicated to the overthrow of established religion and monarchy. The Illuminati then fostered their political
mission by gradually infiltrating the German Masonic movement. Always
on the lookout for new recruits, the officials in Frankfurt were pleased to
welcome the rising Masonic star Count Cagliostro.

Cagliostro was taken to a house three miles from Frankfurt and led into
a secret basement dug in the garden. There, at the bottom of the stairs, dimly
lit by torchlight, he was formally inducted into the Illuminati. On a manuscript swearing the destruction of all tyrants he scrawled the name Alessandro di Cagliostro. It was the twelfth signature, he claimed, on a list that
comprised the grand masters of the Templars. Cagliostro had turned Catherine's phantoms into flesh.

· 4 ·

Copt

Copt: *A native Egyptian Christian belonging to the Jacobite sect of Monophysites.*

AROUND 10 P.M. on 31 May 1785, ten thousand people pushed up the great staircase of the Palais de Justice in Paris and against the railings at the side of the Pont au Change, spilling out from the courts of the Ile de la Cité all the way down to the embankment of the Seine River. The reason for the crush was their feverish eagerness to hear the parlement of Paris, in its judicial capacity as the king's Great Chamber, pass a verdict on the most sensational trial of the eighteenth century, the "affair of the diamond necklace."

People say the mob came from every quarter of the city, pouring across the five bridges of the Seine from Pont Neuf to Notre Dame: fishwives, soldiers, police spies, students, lawyers, shopkeepers, financiers, clergymen, pimps, prostitutes. There were scribblers from the coffeehouses, gamblers from the casinos of the Palais Royal, and a scattering of well-dressed women wearing the favors of Cardinal Prince de Rohan on their bonnets and blouses. The *cardinal sur la paille*, as they called this cluster of red and yellow ribbons, marked the ladies' outrage that a great noble should be lying on a pallet of straw in a cell of the Bastille at the whim of the Bourbon monarchy.

To four people that evening the verdict mattered as much as life itself. One was the cardinal. That such a man should have been tossed into the Bastille seemed unthinkable. His titles alone signified his power: Louis-René-Édouard de Rohan, Cardinal of the Holy Roman Church, Bishop Prince of Strasbourg, Prince of Hildesheim, Landgrave of Alsace, Grand Almoner of France, Commander of the Holy Ghost, Commendator of the Saint Waast d'Arras, Superior-General of the Royal Hospital of Quinze-Vingts, Abbé of the Chaise-Dieu and Master of the Sorbonne. Flaunting the arrogant motto "King I am not, prince I disdain to be, Rohan I am," his family took second place only to the Bourbons in rank and influence. And the Cardinal was their most glittering representative. Nicknamed "La Belle Éminence" because of his playboy ways, the fifty-one-year-old clergyman had an annual income of around 1,200,000 francs, enough to run a small country.

He stood accused of acquiring and stealing a priceless diamond necklace under the forged signature "Marie-Antoinette, Queen of France." Worse, he'd also written the queen a batch of sizzling love letters and gloated over the steamy replies, not realizing they were forgeries. On one occasion he'd literally thrown himself at the feet of a prostitute whom he imagined to be Marie-Antoinette. His shocking presumption that the queen would be willing to receive such advances entailed the crime of lèse-majesté, a near sacrilege for which the penalty was death.

Rohan claimed that a plausible woman imposter, Countess Jeanne de Valois de La Motte, had conned him into believing she was an intimate of the queen. She'd then masterminded the whole swindle by forging letters purporting to come from the queen; she'd persuaded him to buy the necklace on behalf of Marie-Antoinette, and, eventually, Jeanne herself had stolen and sold the necklace. Fortunately for Rohan, lack of concrete evidence against him had forced his prosecutors to ask for a series of noncapital punishments. These included a further spell in the Bastille, an abject apology, loss of all titles, and life-long banishment from France. To the broken cardinal, who'd already been imprisoned since August of the previous year, death was almost preferable.

On the outskirts of Paris, at Versailles, in Europe's largest and grandest palace, a second person waited fretfully that evening to hear the verdict. She was the wronged queen, Marie-Antoinette. Ironically, she'd despised this libertine clergyman for more than a decade, ever since Rohan had relayed some malicious gossip to her censorious Austrian mother, Empress Maria Theresa. While working in Vienna, Rohan had also sullied his office as French ambassador by a passion for gambling, hunting, and fornication. Marie-Antoinette had hoped to strip him of all his titles when her husband became king in 1774, but Rohan's family had proved too powerful and Louis XVI too weak. The best Marie-Antoinette had been able to accomplish was his replacement as ambassador by Baron de Breteuil, a hated rival. She had, of course, isolated the cardinal utterly from fashionable court life.

So when Marie-Antoinette learned in August 1784 that Rohan had used her name to purchase a great ugly diamond necklace worth the colossal sum of 1,600,000 livres, the queen had been stunned, particularly because it had been originally designed for a courtesan. Her mood turned to fury when Rohan told her husband that she'd written him personal letters. This disgusting creature, whom she'd pointedly shunned for so long, later compounded his impudence by saying that he'd genuinely believed a prostitute, whose feet he had slobbered over, to be his sovereign queen, Marie-Antoinette.

From the queen's perspective, it seemed obvious that the cardinal had orchestrated the whole fraud to pay for his notorious debts; she would have had the wretch executed on the spot had France's wishy-washy laws not forbidden it. Instead, with the support of her favorite minister, Baron de Breteuil, she badgered the king into having Rohan arrested under the most humiliating circumstances—in his full robes while preparing to celebrate a royal mass on the Feast of the Virgin, 15 August 1785. He was marched past gawking crowds to the Bastille, to be tried later by the parlement of Paris. "Such hideous vice must be unmasked," Marie-Antoinette angrily told her lady-in-waiting Madame Campan; "when the Roman purple and title of prince are used to cloak a common thief so desperate for money that he dares to compromise the wife of his sovereign, then all France, all Europe, must hear about it."

Count Alessandro di Cagliostro was the third figure on that night of 31 May with an urgent investment in the outcome of the trial. If the cardinal were found guilty, Cagliostro would go down with him, for they stood accused as co-conspirators. Some newspapers claimed that Cagliostro was the puppeteer behind the whole plot: he'd supposedly bewitched the cardinal into believing in the fantasy of the queen's love and persuaded him to steal the diamond necklace.

Cagliostro's key accuser, Countess Jeanne de Valois de La Motte, was the fourth person desperately awaiting a verdict that evening, for she too was a state prisoner charged with a role in the extraordinary affair. This beautiful and savage young woman of twenty-seven boasted royal descent via the ancient bastard line of Valois, long regarded as extinct. Marie-Antoinette, in a letter to her brother Joseph II, emperor of Austria, painted Jeanne as "an *intrigante* of the basest order, although not without a measure of attraction and *ton*, a dashing sort of a creature." Even so, the queen thought her a pawn of the cardinal and Cagliostro. If these two were found guilty, Jeanne would probably be freed as an innocent scapegoat, but if they were acquitted the needle of blame would swing back to her.

And so it did. Jeanne was sitting in a back room of the Conciergerie, a squat medieval fortress overlooking the Palais de Justice, when she heard explosive cheers that signaled the cardinal's triumph. Before fainting into a chair she dimly heard the question, "What will become of poor Madame La Motte?" Hubert, her jailer, carried the prostrate prisoner to a bed, then gently told her that both the cardinal and Cagliostro had been acquitted unconditionally. Knowing Jeanne's ferocious temper, he tried to placate her by suggesting that she herself would probably be confined to a convent for just a few years.

But she was inconsolable, enraged, and proceeded to beat herself bloody with a Dutch china mug. Her forehead split open and blood spurted over her face and clothes. The kindly keeper and his wife had to restrain Jeanne from damaging herself further as her body jerked and twisted in fierce spasms. Hubert was later thankful that he didn't then know the court's real verdict—that "Jeanne Valois de Saint-Rémy [La Motte] have a halter placed on her neck, be stripped naked and beaten with rods, have both her shoulders branded with a hot iron carrying the letter V for voleuse or thief, and be incarcerated for life in the women's prison known as the Salpêtrière."

The cardinal heard of his acquittal shortly after ten o'clock. Pulverized by the ordeal of the past eleven months, the once debonair clerical prince was sitting numbly in the record office of the Palais de Justice. Comte de Launay, governor of the Bastille, tried to stop Rohan's attorney, Monsieur Target, from entering the room, until the lawyer barked that the prince de Rohan was now a free man. Even so, Rohan would have one last night to spend in the Bastille before his official release.

Marie-Antoinette, pacing up and down her boudoir at Versailles, was aghast to hear that the odious cardinal had escaped his just desserts. She scribbled a tearstained note to her lady-in-waiting Juliet de Polignac, begging her to "come and weep with me, come and console your friend.... The judgment which has just been pronounced is a shameful insult. I am bathed in tears of grief and despair." Another letter, written to her sister in Austria on the same night, raged at the humiliation of allowing "a perjured priest, a lewd intriguer" to go free when he had committed high treason against his sovereign's wife. Madame Campan, a lady-in-waiting on the spot, heard Marie-Antoinette spit out the accusation that most of the court had been corrupted by the Rohan family or had simply seized the chance to hurt the

Bourbons "in so marked a manner." Looking back on this moment, Madame Campan saw it as the prelude to the end: neither Marie-Antoinette nor France would ever really recover.

Count Alessandro di Cagliostro was the only one of these four to turn the ordeal into a triumph. Even on the way to court in the morning, he'd entertained the crowd with a showman's antics. Dressed in a green velvet coat embroidered with gold lace, he'd raised his arms above his head and thrown his hat in the air to show a bizarre hairstyle that began with a tight braid on top of his head then tumbled down his shoulders in wavelets of tiny curls. When he was released from the Bastille at around eleven the next evening, he was mobbed by cheering citizens. Street vendors did a brisk business selling engravings of his portrait, while a cripple scurried alongside his carriage, handing out free samples of the famous "wine of Egypt" and other healing potions.

When Cagliostro's black japanned coach finally rumbled into his courtyard on rue St-Claude, off the boulevard St-Antoine, "the night was dark, the quarter in which I resided but little frequented. What was my surprise, then, to hear myself acclaimed by eight or ten thousand persons. My door was forced open; the courtyard, the staircase, the rooms were crowded with people. At such a moment my heart could not contain all the feelings which strove for mastery in it. My knees gave way beneath me. I fell on the floor unconscious. With a shriek my wife sank in a swoon. Our friends pressed around us, uncertain whether the most beautiful moment in our life would not be the last.... I recovered. A torrent of tears streamed down from my eyes."

No one has managed to penetrate all the murky layers of the "diamond necklace affair," but the contemporary figure who came closest was Abbé Georgel, a Jesuit priest and man of business who served as Rohan's secretary for more than twenty years. To the last, Georgel remained convinced that Cagliostro "beyond any man alive, held the key to the enigma: knowledge of the secret motives behind it all." It was this diabolical conjuror, Georgel argued, who began the fateful chain of events by bewitching the cardinal in Strasbourg in 1781. "I do not know," he reflected bitterly, "what monster of evil, what archenemy of mankind vomited upon these shores this scintillating new genus of charlatanry, this new apostle of the 'universal religion' who wielded despotic power over his proselytes, subjugating them entirely to his will." Georgel's master Rohan had been among those Cagliostro "seduced into the treacherous bypaths of the occult and supernatural."

Georgel was certain that the conjuror had come to Strasbourg on 19 September 1780 — immediately after leaving Poland — with the ultimate intention of conning the cardinal. Cagliostro had been invited to the city by senior Masonic officials of the Strict Observance Rite and expected an enthusiastic reception. Though Strasbourg had a population of only 50,000, its situation on the French-German border and its mixed French-German population made it a hub of European Freemasonry. At the time of Cagliostro's arrival there were twenty-nine operative lodges with a total of around 1,500 members.

In spite of receiving a warm welcome from crowds of well-wishers who knew his reputation for healing the poor, Cagliostro had been disconcerted to find that most of Strasbourg's Masonic community avoided him. Even Strict Observance Masons grew wary when they discovered that Count Cagliostro was presenting himself no longer as an agent of their rite but as founder and leader of Egyptian Freemasonry. He'd begun his mission to Russia and Poland still claiming to be an emissary of the "Unknown Superiors" — including the "Great Copt," an immortal high priest who'd supposedly fathered Egyptian Masonry at the Pyramids in ancient times. Somewhere en route to Strasbourg, however, Count Cagliostro had elevated himself. He'd graduated from working for the Great Copt into being the Great Copt himself. And though he didn't actually claim to be immortal, he hinted broadly that he was.

Strasbourg didn't at first live up to its reputation. Its Masonic scene was unusually snooty and fractious, a reflection of the troubled recent history of the city. Once a Protestant and German-speaking free city-state within the Holy Roman Empire, it had been grafted onto the Catholic autocratic empire of France during the seventeenth century. As a result, social divisions ran deep. Lodges bickered interminably: Catholic against Protestant, bourgeois against aristocrat, French against German, reformist against conservative, military against bureaucrat.

As in Saint Petersburg, Cagliostro responded to social exclusion by turning to the common people. For the next twelve months he threw himself into a mission to heal and succor the poor of Strasbourg. His clinic was besieged by the crippled, deaf, blind, and indigent, and his carriage was noticed outside hovels in the poorest quarters of town at all hours of the day and night. "The most interesting thing in this Gallico-German town," observed Farmer-General Jean-Benjamin de La Borde, who met him in Strasbourg, "is the kind of audience held by the Count Cagliostro. . . . He is adored by the poor and *petit peuple*; also hated, reviled and persecuted by certain other people. He receives neither money nor gifts from those he cures. He spends his life among the sick, above all with the poor, distributing his remedies free of charge and paying out of his own pocket for soup. He eats little himself, and that almost always pasta from Italy, never goes to bed, and sleeps only two or three hours in a chair. He is always ready to fly to the aid of the unhappy no matter what hour of the day or night and he has no other pleasure in life but to sooth his fellow-beings."

Soon wealthier patients also began to seek his services, especially patients with supposedly incurable diseases. He made his first social advance in the military garrison, an influential sector within a border city that protected a strategic crossing point of the Rhine. After Cagliostro cured one military man of gangrene and another of suicidal depression, his fame spread rapidly. Encouraged by Marshal Contades, commander in chief of Alsace, officers welcomed the Copt to their messes and lodges. Wives and sweethearts followed suit.

The Marquise de la Salle and Baroness Flachsland formed the nucleus of a salon that began to gather every evening at the Cagliostros' fine new premises situated on the corner of the place d'Armes, Kalbsgasse 1, known as the "House of the Virgin." Here Seraphina presided in decorous silence until Cagliostro, fresh from his clinic, charged into the room like a stocky bullock. He would stand, back to the fire, stomach and chest thrust out, enthralling the company with stories of his Arabian travels.

Eventually the reputation of this exotic salon reached Cardinal Rohan's palace at Saverne on the outskirts of Strasbourg. To Rohan, living away from Paris felt like exile. He found the Alsatian city irritatingly straitlaced and dull. Having inherited the position of archbishop from his uncle, he felt no obligation to perform any duties beyond collecting the rich annual fees. More than half the population of the province was Lutheran anyway. Rohan pined to sip the nectar of love and power around Versailles, but Marie-Antoinette had banned him from its blossoms. To alleviate the boredom, he sent word through his master of the hunt that he wished to call on this Count Cagliostro. To his surprise, he received the arrogant reply that the Great Copt would see only the sick or needy. A masterstroke, this: Cagliostro's playing hard to get inflamed Rohan's interest. Instantly the cardinal developed a bout of asthma and presented himself for treatment at Cagliostro's door.

From that moment, Georgel believed, the cardinal was hooked. He returned from his first meeting with Cagliostro as excited as a puppy, telling his skeptical secretary that he had seen deep within the "Great Copt's" face "a dignity so impressive that he felt himself in the grip of some awesome religious experience." For quite other reasons he seemed equally excited by Countess Cagliostro, who flirted with him as though he were a young beau. Under this dual influence, Rohan's asthma vanished. Before long, the cardinal earned reciprocal approval. "Your soul," Cagliostro condescended, "is worthy of my own. You deserve to be the confidant of my secrets." What greater honor could any man hope for?

One of these secrets was needed urgently in July 1781 to save the cardinal's military uncle, the prince de Soubise, who was reported to be dying of a fever. Rohan and his new guru dashed to Paris to mount a rescue mission. The business was delicate because Parisian physicians would not take kindly to the intervention of a strange Arabian healer, so Cagliostro visited the prince in disguise, administering his drops and setting out an accurate timetable for the patient's return to health. Sure enough, Prince Soubise recovered exactly as predicted.

After this, Georgel says, the Copt became Rohan's "oracle, his guide, his compass." Cagliostro began by steering the cardinal back to an earlier interest in chemical experimentation. Some years earlier Rohan had dreamed of making money by turning salt water into saltpeter; now, dressed in a wizard's cloak and pointed hat, he worked as Cagliostro's laboratory assistant, helping him grow diamonds and transform base matter into golden ore.

Baroness Henriette-Louise d'Oberkirch, a fine-boned Protestant aristocrat from upper Alsace, visited Rohan's Saverne Palace twice during the winter of 1780–1781 and observed Cagliostro's growing power over the cardinal. She was honest enough to admit that the Great Copt almost succeeded in bewitching her too. Though determined to resist the man's allure, she found herself transfixed by his eyes: "They were indescribable, with supernatural depths—all fire and yet all ice." She felt them boring into her brain like a drill. His voice caressed her, "like a trumpet veiled in crepe." His haughty manner, she noted later, "at once attracted and repulsed you, he frightened you and at the same time inspired you with an insatiable curiosity." The most startling moment was when he suddenly interrupted the conversation, as if struck with a sudden vision, to announce that Rohan's enemy Empress Maria Theresa of Austria had that moment died. Cagliostro turned out to be right, though no one else in Strasbourg heard the news until five days later.

On her next visit in January 1781 Henriette-Louise felt her heart hammering with excitement as she entered the dining room of the palace. She scorned the way local women fawned on Cagliostro with "an infatuation little short of idolatry," but she had to struggle not to go the same way. More than anything, she longed to question him about her personal destiny. At dinner, the cardinal went out of his way to trumpet Cagliostro's miraculous powers. He flourished a gorgeous diamond ring, valued at 25,000 francs, which he said Cagliostro had grown in an alchemical oven before the cardinal's very eyes. The Copt had pressed the cardinal to accept the precious gift, airily refusing any payment. "He will make me the richest man in Europe," Rohan boasted. Henriette-Louise guessed that Cagliostro was trying to impress her as a way of getting to Catherine the Great, who was her close friend. Even so, the baroness couldn't help being fascinated: "Cagliostro was possessed of a demonic power, he enthralled the mind, paralyzed the will."

Abbé Georgel, too, noticed that Rohan had fallen completely under Cagliostro's spell. When called away to Paris on business, the cardinal pressed the Cagliostros to stay in his palace so that they could be pampered by his Swiss factotum, Baron de Planta. Georgel was indignant to think of wondrous Saverne in the hands of these scumbags. Georgel told his master that the palace's best wine had flowed like water in his absence, but the cardinal only laughed indulgently, saying that Cagliostro deserved some good times.

Good times, indeed! Giuseppe Balsamo, issuing daily orders to the twenty-five footmen and fourteen *maîtres d'hôtel* employed in Rohan's vast sandstone palace, could pride himself that he'd come a long way from the slums of Palermo. Looking from the front door over the valley of the Rhine, he took in a domain that included two towns, a cathedral, and an ancient château. Come, Baron Planta, let us drink up. Let the Tokay flow.

Despite Georgel's certainty that Cagliostro had come to Strasbourg only to rob his master, there was plenty of evidence to the contrary. Money flowed to Cagliostro more easily than Tokay. In the summer of 1781, for example, he performed what turned out to be the most lucrative healing of his life. A fabulously rich banker and silk merchant, Jacques Sarasin, brought his wife from the Swiss town of Bâle to Cagliostro's clinic in Strasbourg. Sarasin was desperate; he'd tried every other medical and spiritual recourse without success.

Gertrude Sarasin was gripped by a mysterious wasting fever. Often she slept only an hour a night. When she was seized by convulsions, eight people were needed to hold her down. A regimen of Cagliostro's treatments produced immediate results. The fever abated, and she began to eat and sleep like any normal person. Within a few months of her completion of the treatment, Jacques was able to write to Cagliostro from Switzerland that his beloved Gertrude had even conceived again. Naturally they named the child Alexander after their savior. The relieved banker poured out his testimonial in a Parisian newspaper: "To express my gratitude would be very difficult: I do not have the words to define the feelings of my heart." Jacques would come to Cagliostro's rescue in a later chapter of his life; in the meantime he did what he could to repay the Copt by providing him with unlimited bank credit obtainable anywhere in Europe.

Far from battening on Rohan like the bloodsucking leech of Georgel's account, Cagliostro distressed the cardinal in June 1783 by announcing that after three years of philanthropic medical practice, it was time to move on. Working in Strasbourg was becoming difficult because he'd antagonized an influential local physician by healing one of his patients. Such presumptuous behavior provoked the city's medical fraternity into a campaign to drive the quack out of town. They paid one of Cagliostro's former employees to smear him in newspapers and tracts. This disgruntled druggist, Carlo Saachi, said he'd treated Cagliostro for venereal disease in Spain and was owed back wages. He didn't mention that he'd been dismissed for selling Cagliostro's recipes on the side. Most, if not all, of his accusations were invented, but they were music to the ears of the doctors in Strasbourg.

Cagliostro decided to head for the south of France—the heartland of mystical Freemasonry where precursors like Saint-Germain and Swedenborg had established large followings. Cardinal Rohan, desolate at the prospect

of losing contact with his spiritual master, appointed
a young scientist, Raymond de Carbonnières,
to look after the Cagliostros and maintain a
lifeline between them and himself.

Much as Georgel disliked the Great Copt,
he admitted in his memoir that Cagliostro
had been rivaled at this time by another
diabolical figure who also wanted to bewitch
the cardinal. Georgel nicknamed this woman
Circe — the evil sorceress of Greek legend who
lured men to their fate. Others knew her as Countess Jeanne de Valois de Saint-Rémy de La Motte.

Jeanne de La Motte

There were strange affinities between Rohan's
two enchanters. Jeanne, like young Giuseppe, had grown up in poverty
and honed her wits on the streets. Her family, like the Balsamos, talked nos-
talgically about their noble bloodlines. In Jeanne's case, this became an obsession. She'd been born in 1756 at the chateau of Fontette just outside the old
walled village of Bar-sur-Aube in Champagne. Her father, Jacques de
Saint-Rémy, had once carried the title Baron of Luze and Valois — as a
direct descendant of a bastard royal line going back to Henri II. Little
Jeanne grew up nursing a fantasy: in her mind she was always an outcast
princess.

In reality, the girl knew her father only as a wine-soaked derelict married to the daughter of a lodge keeper who'd once worked for him. The
baron of Valois lived by poaching and stealing from the estates he'd sold to
pay for his gambling debts. Jeanne's first house — a wooden hut fitted with
a small trapdoor through which locals would push scraps for the children — made Giuseppe's birthplace seem palatial. Around 1760, her family
drifted to Paris to look for work; but within a year her father died in the

Hôtel-Dieu poorhouse, leaving Jeanne at the mercy of her mother's new lover, a brutish Sardinian guardsman. As well as raping his eight-year-old stepdaughter, he made her the chief family earner; she was sent out daily to beg in the streets around Boulogne.

If Giuseppe acquired his flair for magic in the markets of the Albergheria, Jeanne's new habitat taught her the arts of enticement. She learned to exploit every atom of her sad story. One brisk March day in 1764, for example, the marquise de Boulainvilliers stopped her carriage outside the village of Passy near Boulogne to speak to a thin shivering girl with her baby sister strapped to her back. She was holding a sign, "Pity a poor orphan of the blood of Valois." From that moment the benevolent marquise became both patron and victim of this feral child with a sweet smile and a will of iron. Over the next decade the marquise had Jeanne and her sister trained as mantua makers; then she tried to civilize them in a convent for poor gentlefolk. Meantime, the marquis de Boulainvilliers taught Jeanne a different set of skills on nightly visits to her chamber.

At the age of twenty, under pressure to take holy vows, Jeanne ran away from the convent to her hometown of Bar-sur-Aube and there threw herself on the charity of Madame Surmont, the kindly wife of a local judge. A year later, Jeanne's benefactress was complaining that she'd nurtured "a demon": the girl had bewitched Monsieur Surmont, become pregnant by the local bishop of Langres, then ensnared madame's nephew, a dissolute young guardsman named Nicolas La Motte. A hastily arranged marriage with Nicolas on 6 June 1780 covered up the bishop's indiscretion — twin boys who conveniently died within a few weeks of their birth.

A lawyer, Albert Beugnot, another young man of the district who fell in love with Jeanne at this time, described her strange sexual allure: "Madame de La Motte was not what you would call a beauty; she was of average height but lissom and well-proportioned; she had very expressive blue eyes, under dark brows, strongly arched; her face was a trifle long, her mouth wide, showing beautiful teeth; and what is perfect for such a type, her smile was enchanting. She had beautiful hands, tiny feet. Her complexion was of a remarkable whiteness.... She was devoid of all learning but she had a plenty of intelligence and quick penetrating wit." Reacting—as another of her lovers said—"to every flare of passion," she could also pursue her goals with icy calculation and remorseless determination. "I have read authors who assert that courage is...the characteristic of the male sex," she wrote in the second of her autobiographies. "...I have conceived it a desirable qualification, which though a female, I have endeavored to attain." She liked to disguise herself by dressing in men's clothes; she possessed huge physical stamina; and she could fight with maniacal ferocity.

Jeanne's two longest-serving lovers—grave, dark Albert Beugnot and blond, lusty Rétaux de la Villette, a guardsman friend of her husband—doubted that she ever loved or liked anyone at all. Beugnot considered her a type of *enfant sauvage*: "Engaged in perpetual conflict with society from the time of her birth, she had learned to disdain its laws, and had but little respect for those of morality." Giuseppe Balsamo was thirty-five when he discovered his life's mission of Freemasonry; Jeanne knew hers almost from birth. The only thing that made her pain, humiliation, and hunger endurable was a determination to retrieve the lost lands of her Valois father. This *idée fixe*, Beugnot said, made her "a woman who would stop at nothing to achieve her ends and who felt entirely justified to defy a social order which had denied her birthright."

Jeanne's first meeting with Cardinal Rohan around September 1781 reflected all this. Hearing that Madame Boulainvilliers was visiting the famously rich cardinal in Strasbourg, Jeanne rushed there from the Lunéville barracks sixty miles away. She contrived to bump into Rohan as he was taking a drive in his carriage. It was worth the effort. Jeanne had the perfect blend of ingredients: a sob story to touch his sentimental nature, hauteur to engage his snobbery, and sensuality to arouse his notorious lust. As grand almoner of France, the cardinal was required to help the hard-up nobility on behalf of the king; he did his duty by handing Jeanne a fat purse with instructions to contact him again if she was ever in need.

She was always in need. But her next meeting with the cardinal, in the summer of 1782, required more elaborate planning. First, she moved to Paris and took furnished rooms on the top floor of a hotel in the Marais, conveniently close to Rohan's grand palace. In the interim she'd also managed to become a "countess," simply by lifting the title from an unrelated branch of La Mottes. She recruited young Beugnot—who'd also moved to Paris, supposedly to further his law career—to compose endless pleas for the reinstatement of her lands in exchange for occasional access to her bed. Having found out that the cardinal was staying in Paris, Jeanne, with typical cheek, demanded to borrow Beugnot's carriage and servant in preparation for a visit to the Palais-Cardinal.

What a spot for a seduction. As she entered the palace through the marble vestibule, Jeanne had to pass four suggestive Boucher paintings of Mars reposing with Venus. A servant ushered her upstairs into the *salon des singes*, where the cardinal conducted private mass. A strange chapel, this: around the wall, lacquer-colored paintings by Robert Huet showed fleshy nobles in scanty Chinese costumes. But these looked tame beside a series of miniature murals on the timber-paneled walls and altarpiece, each depicting a scarlet-coated simian in an obscene posture. One monkey grinned lasciviously as it snuffed out a candle with its anus. The cardinal, with a gesture, invited Jeanne onto the couch next to him and asked to hear her sad story again. He pressed her hand to convey his sorrow that she, a woman "fashioned expressly by nature for the conquest of men's hearts," should have so useless a husband. Soon he sprang a hidden door in the altar paneling that opened to reveal a boudoir more reminiscent of a courtesan than a cardinal. The chivalrous Rohan also showed her a staircase running directly into the gardens below so that ladies could leave with complete discretion.

The next day Jeanne was exultant; within a few days she told Beugnot that she no longer needed his services as lawyer or lover. With the cardinal as her "protector," she would try a new plan. She'd locate herself in Versailles, fourteen miles southwest of Paris, and lobby the court directly for restoration of her Valois titles and lands. Poor discarded Beugnot warned her to be careful: Versailles was infested with swindlers and professional intrigants, "always on the verge of making a fortune, always glib with dangerous advice about how to go about things which should not be done."

It's hard to imagine a more fantastic scheme than hers. How could this penniless adventuress hope to storm the most exclusive, protocol-bound court in Europe? True, Jeanne did have two things going for her. First, Versailles was less a palace than a type of courtier state, so huge and amorphous that it had produced its own parasitic underclass. A clever imposter might insinuate herself among the 12,000 people who serviced the building each day. She could walk confidently along the miles of dim corridors, mingling with hurrying officials, servants, and workers of all kinds, or she could stroll brazenly through the acres of surrounding gardens so packed with hangers-on that the smell of their excreta drifted up to the queen's windows.

Jeanne's second advantage was that few other intrigants could possibly match her for wit, courage, and ruthlessness. Even so, the next few years tested these capacities to the limit. After taking rooms at the Belle Image Inn in Versailles, she bombarded officials and members of the royal family with requests for pensions, property, and titles. The controller general of finance, Charles Alexandre de Calonne, eventually granted her an increase in pension to make her go away. One day she managed to sneak into the reception room of the king's sister, Princess Elisabeth, where she threw herself to the ground in a fainting fit. Claims that she was malnourished and ill from a miscarriage won her more royal charity, but she destroyed this sympathy by trying to repeat the stunt a few months later.

Banned from the palace's inner purlieus, she prowled around its edges, soaking up gossip and courtier's slang and buttering up menials like Goubert, the gatekeeper of the queen's favorite garden at the Petit Trianon. When

Jeanne could beg no more from Rohan, she gave favors to noblemen who visited her dingy hotel room, in exchange for gifts and money. Sometimes she was literally famished. The long-suffering Beugnot took her for a cheap meal occasionally and was shocked at how greedily she gulped her beer and wolfed down several dozen little cakes at a sitting.

In defiance of her deepening poverty, Jeanne decided to put on an appearance of gentility by moving back to Paris and renting, on credit, an elegant three-story house with iron balconies, a porter's lodge, and stables. It was perfectly situated at 13 rue Neuve-Saint-Gilles, in the Marais, five minutes from the cardinal's palace. She didn't have far to travel at night, but the house, like so much else in Jeanne's life, was largely an illusion. Most of the time she lived in a single upper room with her servants, storing her furniture at a nearby barber's shop so as to evade creditors. Her servant Rosalie worked for months without wages and sometimes paid Jeanne's most pressing debts from her own savings.

But Jeanne never wasted an opportunity. An ingenuous little priest, Father Roth, from the nearby Minimes chapel was so moved by her plight that he became a type of unpaid secretary. Her earlier lover from Bar-sur-Aube, Rétaux de la Villette, also reappeared, an excellent addition to Jeanne's shabby court. Rétaux could forge handwriting, compose verses, sing, and, by his own account, make love like a bull—skills that could prove useful to an imposter countess.

The first breakthrough during this difficult time came when several hopeful tradesmen approached Jeanne to act as a royal lobbyist for them: they'd heard stories that she was on close terms with powerful figures at court, including the queen. Jeanne had been fanning these rumors for some time, and at last the flame had caught. In April 1784 she decided to try out a version of this lie on the cardinal himself. Rohan's joy at learning of her supposed friendship with the queen told Jeanne that she'd hit the jackpot. The cardinal longed to restore his damaged reputation with Marie-Antoinette: only her rancor, he believed, stopped him from becoming first minister of

France. This vain fantasy was—Georgel said—the "infernal spark" that ignited the diamond necklace affair. And of course it was Jeanne who struck the tinder.

Throughout the long and delicate entrapment of the cardinal she never made a false step. Using a combination of inside knowledge and psychological insight, she caught Rohan in what Beugnot called a "cloak of whispers." Jeanne's intelligence, Rétaux's penmanship, and some costly gilt paper with flowered borders were the ingredients that created an imposter queen. Jeanne began by urging the cardinal to write to Marie-Antoinette offering an explanation and apology for his past misdeeds. A brilliant idea: Rohan, eager as a spaniel, redrafted this letter some twenty times before sending Jeanne off with it.

Little by little Jeanne moved her faked replies from icy formality to intimacy, as she allowed the queen to succumb to the cardinal's oily excuses and suggestive words. In concocting these letters, Jeanne brilliantly exploited Rohan's vanity and Marie-Antoinette's undeserved reputation for promiscuity. During a long period of emotional frustration at the beginning of the reign, when a minor anatomical problem had prevented the king from consummating their marriage, Marie-Antoinette gained the reputation of a "pretty featherhead." These were the words of her brother Joseph II of Austria, but plenty of people at court called her worse names. Offended by her silly friends and her cavalier ways, they whispered that she was a predatory lesbian and an insatiable lover of men. The cardinal, who saw the world entirely through a libertine's lens, thought it perfectly natural that the queen should fall for a beauty like Jeanne and a beau like himself.

As time passed, though, Rohan chafed for a personal audience with Marie-Antoinette; he longed to move the affair onto a physical plane. A tricky moment, this, but it was now that Jeanne's cloddish husband Nicolas had his one inspirational moment. He was cruising the gardens of the Palais-Royal one midsummer afternoon when his roving eye fell on an attractive twenty-three-year-old called Nicole Leguay. These former palace gardens of the king's cousin had been redeveloped into arcades of business and pleasure, crammed with booths, casinos, cafés, and printing shops. Being free from police interference because of their royal status, they'd become a rich humus for Parisian vice.

Nicole was a typical inhabitant of the Palais-Royal, much as Jeanne would have been without the Valois pretensions and a streak of genius. She was a pretty milliner and part-time prostitute with little education but a burnish that comes from entertaining nobility. Sharp-eyed Nicolas also noticed something else about her: Nicole's soft blond hair, cornflower-blue eyes, and long pale face looked uncannily like portraits of Marie-Antoinette. Quickly he ingratiated himself with the charming grisette, then rushed to pass the news to his wife. And so the plan was hatched to make Nicole Leguay queen for a night.

Jeanne was the impresario who elevated this mad idea into one of those scenes where life outbids art. Drawing on a famous portrait of Marie-Antoinette by Madame Vigée-Lebrun printed on the back of a jewelry box, Jeanne used her milliner's skills to turn Nicole into the queen, "wearing a simple white dress, without any ornament on her head other than her hair dressed in the fashion of the time and the two ringlets which, on either side, fell to her neck."

From the plays of Beaumarchais came the Figaro-style plot of having a prostitute impersonate a queen to deceive an amorous bishop. Jeanne's own theatrical genius produced the perfect mise-en-scène—a midnight tryst on a hot, moonless summer night. The setting was the the "grove of Venus," a leafy grotto in the queen's favorite Petit Trianon garden. Nicole, who'd been accorded a dazzling new name, Baroness d'Oliva (a rough inversion of Valois), believed that she was being asked to play a little joke on behalf of the real queen, who would be peeking from behind the foliage. Meantime, Cardinal Rohan, disguised in cloak and broad-brimmed hat, had whipped himself into a fever of excitement at the prospect of this dangerous liaison.

If the cardinal had any residual doubts about the relationship, they melted with his heart when Nicole muttered a few words, thrust a red rose into his hand, and hurried into the night. Jeanne's first memoir contains a plausibly ecstatic letter that Rohan supposedly wrote to the queen the following day: "That charming rose lies upon my heart—I will preserve it to my latest breath. It will incessantly recall to me the first instant of my happiness.... I was so transported, that I found myself imperceptibly brought to the charming spot which you had made choice of. After having crossed the shrubbery, I almost despaired of knowing again the spot where your beloved Slave threw himself at your feet.... I found again the pleasing turf, gently pressed by those pretty little feet—I rushed upon it, as if you had still been there, and kissed with as much ardor your grassy seat."

At first, Jeanne intended simply to bleed the cardinal for as long as the deception could be sustained. She immediately began requesting money in the queen's name, always for some charitable purpose, of course. Bad luck that the cardinal had in turn to borrow the sums from a Jewish moneylender. By some odd coincidence, Jeanne's standard of living took a sudden turn for the better. As for Nicole Leguay, she was rapidly discarded after being paid less than half the amount promised for giving the performance of her life.

Then, in December 1784, came one of those strokes of fortune that bless the truly bold. The court jewelers, Bassenge and Bohmer, anxious to sell a fabulously expensive diamond necklace that was locking up their capital, approached Countess La Motte to lobby "her friend, the queen" to buy this glorious set of jewels. With lightning acuity Jeanne saw her chance. Via the usual blue-flowered letter, Rohan would be asked to underwrite the sale of the necklace on behalf of the queen. A temporary loan, you understand—Her Majesty was a little hard-pressed for cash at the moment and didn't want to annoy the king with evidence of extravagance. Of course absolute discretion was necessary, but the queen would not forget such a favor.

Here again the delicate orchestration of the scam required steel nerves and brilliant acting. Though overjoyed at having won the queen's heart, the cardinal was still jumpy. The terms of the proposed sale were steep, and the political stakes enormous, even for someone as rich and ambitious as himself. After intense negotiation, terms of agreement were reached on 29 January 1785. The price was to be a colossal 1,600,000 livres, payable in quarterly installments over two years. The necklace would be handed over on 1 February.

Two days before the delivery, Jeanne returned from Versailles, claiming that the queen had agreed to the sale but did not wish to put her signature to any document. To Jeanne's alarm, Rohan proved less compliant than usual, insisting on a royal signature. Then suddenly he lobbed a second grenade: he'd persuaded his divine master, Count Cagliostro, to come to Paris. In fact, the Great Copt was arriving at a hotel in the Palais-Royal this very evening. Rohan would consult him about the awesome responsibility of underwriting the necklace. The good spirits would know if it was the right thing to do.

A combination of coolness and luck enabled Jeanne to weather this dangerous moment. To her relief, both the cardinal and the jewelers were so excited that they made no effort to check the authenticity of the signature, "Marie-Antoinette de France," penned clumsily by Rétaux on the side of the bill of sale. Neither did they realize that the queen, as an Austrian patriot, never signed herself this way.

Thank heaven, too, that when the cardinal poured out his heart to Cagliostro that night, the Copt happened to be in a good mood. After taking stock of the spirits, he pronounced the signs favorable: Cardinal Prince Rohan would soon gain the power and influence he desired. Looking back on the affair, Abbé Georgel thought this a key moment. Once the cardinal received Cagliostro's spiritual reassurance, he hesitated no longer—Jeanne's trap was sprung.

On the evening of 1 February, under Jeanne's instructions, the cardinal hid himself in a dim alcove on the first floor of her apartment at place

Dauphine in Versailles and watched the countess solemnly hand over the boxed necklace to a tall, darkly dressed "royal attendant"—in reality, the "stud bull" and forger, Rétaux. As soon as Rohan left, Jeanne, Nicolas, and Rétaux ripped the stones from their settings with a knife and begun to sketch out plans for how and where to sell them.

But selling such numerous and showy diamonds proved difficult; Count La Motte had to take most of them over to England, where suspicions would not be so easily aroused. In the weeks that followed, Jeanne was conscious that the magician could

Count Cagliostro,
bust by Jean-Antoine Houdon

still jeopardize her scheme. Rohan behaved like a deferential child in his presence. He'd even commissioned a marble bust of Cagliostro by the famous French sculptor Houdon, depicting the Copt with his eyes raised to heaven and hair flowing down his neck. Its base carried the words, "These are the features of the friend of mankind. Each day is marked by a new act of goodness. He prolongs life and helps the poor. The pleasure of being useful is his only reward."

Jeanne's own house was now almost literally overshadowed by the far grander *maison de Cagliostro* at 1 rue St.-Claude, off the boulevard St.-Antoine. Rohan had chosen the premises himself so that he could stroll over from his palace most evenings to consult with the seer and flirt with Seraphina, whom he called his "petite comtesse." The costly gifts Rohan heaped on the woman suggested to Jeanne that this was a particularly close relationship. As for the Cagliostros' house: it was fit for royals. It boasted a high-walled portico, two flagged courtyards filled with shady trees, and a long stone staircase winding up three floors to spacious chambers. Jeanne rightly suspected that Rohan had funded the alembics, athanors, creusets, and flasks that bubbled on the stoves every day, as well as the lush Oriental furnishings of the *chambre égyptienne* where the Copt staged his necromantic séances. Here, a black embalmed ibis stood on stiltlike legs, a stuffed alligator with gaping jaws rotated slowly from the ceiling, and strange hieroglyphs covered the walls.

Jeanne could hardly fail to notice that a cult of Cagliostro was sweeping Paris. It was as if the city, so intoxicated by novelty, had been waiting for him to arrive. Cagliostromania was supplanting the balloon craze of the last few years, which had induced Jeanne to buy a ten-foot-high blue cabriolet shaped like Pilastre de Rozier's famous hot air balloon. Given that aeronauts could now soar to the heavens and Dr. Mesmer could make people swoon with magnetic waves, even Cagliostro's fantastic claims seemed possible. After all, his Egyptian Masonry entailed science, religion, and magic.

Princess Guémené, the cardinal's niece, was one of the first Parisian socialites to join a new Isis lodge that Cagliostro and Seraphina established in the rue Verte on 7 August 1785. The Copt's recent stay in the south of France had consolidated his movement's fame, bringing influential new devotees such as the Masonic leader J. B. Willermoz and his clever nephew, Rey de Morande. The latter helped Cagliostro systematize Egyptian Rite ritual into a booklet. This same Lyons circle also funded the rite's first purpose-built mother lodge, known as La Sagesse Triomphante. Egyptian Freemasonry could now keep company with prestigious European rivals. Though Jeanne didn't know it, Cagliostro's real reason for coming to Paris had been to attend a special international congress of mystical Freemasons called to reach an alliance with his movement. The congress had collapsed when Cagliostro declared haughtily that other rites must burn their archives and join him unconditionally.

This condescension was fed by the luster of his new Parisian disciples. The titled nobility flocked to him, including the prince de Montmorency, who accepted the title of Parisian patron of the Egyptian movement; and the duc d'Orleans, Royal Prince and Grand Master of French Masonry since

1771. A small "Supreme Council of the Egyptian Rite" was led by the treasurer-general of the French marine, Claude Baudard de Saint-James, one of Cardinal Rohan's longest-suffering creditors.

Another devoted admirer, Madame de Flammerans, tried unsuccessfully to get her uncle, the bishop of Bruges, to pressure the pope into recognizing Cagliostro's rite. Fashionable Frenchwomen were particularly drawn to Egyptian Masonry because it encouraged them to join mixed lodges that offered a pleasing cocktail of pro-woman rhetoric, glamorous ritual, and sexual titillation. Women initiates, wearing simple white tunics and colored sashes, gathered in an illuminated temple to hear Seraphina urge them to break their chains and force the male sex to beg for favors. Behind her stood a dozen countesses and marchionesses with the title of ancient Roman priestesses or sibyls. Each new initiate was introduced to Seraphina, who breathed on her face to symbolize the strengthening of the soul. Later on, the Copt himself appeared in magnificent red robes, with a serpent rod in one hand. After he, too, promised women Masons equality within the world of the spirits, the evening ended with music, gossip, and drinking.

Jeanne lacked the social credentials to join this elite assembly, but she knew she must win over the arrogant Copt. Cagliostro was one of the few people who even knew of the plan to buy the necklace. If he were to whisper doubts into the cardinal's ear, the swindle could collapse. On the other hand, Cagliostro's approval could strengthen her credibility. Sometime in March, Jeanne whispered to the Copt in confidence that a *very great noblewoman* was worried about her pregnancy and had asked if Cagliostro might predict its outcome. Cagliostro was encouraged to guess that the pregnant woman must be Marie-Antoinette herself. Flattered, he instantly agreed, though he would need a very special *colombe* for so delicate an operation. According to Beugnot, "she had to be of a purity equal only to that of the angels, to be born under a particular astrological sign, to have delicate nerves, great sensibility, and blue eyes." By a happy chance, Jeanne's beautiful teenage niece, Marie-Jeanne La Tour, fitted the bill perfectly.

In front of an elect assembly of some forty people at the cardinal's palace, Cagliostro's assistants set up his usual paraphernalia: wax candles, Masonic insignia, red roses, silken ribbons, pentangles, a double-edged sword, a crystal carafe, and a screen. He performed his initiatory rituals. If you are innocent, Cagliostro told the *colombe*, the good angels Michael and Gabriel will send you a vision; if you are impure, you will see nothing but an ordinary glass. Faced with this tough choice, the girl hesitated. Just as Cagliostro was about to declare her impure, she suddenly saw a distinct image of the queen, pregnant and dressed in white. The good spirits promptly predicted that Marie-Antoinette would give birth to a healthy boy. They turned out to be right, as usual. A boy, Louis Charles, was born on 27 March 1785.

Henceforth, Jeanne made a special effort to charm the Copt. Beugnot described a dinner at her house on rue Neuve-Saint-Gilles, at which she flirted with Cagliostro until he was putty in her hands. Beugnot himself was repelled by the man's "goggle eyes and wide retroussé nose," but impressed by his dress. He wore an iron-gray coat trimmed with gold, a scarlet waistcoat, red knee-breeches, an embroidered hat with a large feather, shoes with jeweled buckles, and sparkling rings on his fingers. The skeptical Beugnot couldn't help being bowled over by the sheer force of the man's rhetoric: "When he attacked a subject he seemed transported, and treated it in the loftiest style by voice and gesture.... I remember nothing of his conversation except that the hero spoke of heaven, the stars, the great arcanum, of Memphis, of hierophancy, transcendental chemistry, giants, of monster animals, a town in the interior of Africa ten times as great as Paris where he had correspondents...and...he varied his discourse with burlesque inanities addressed to Mme de la Motte, whom he called his doe, his gazelle, his swan, his dove." Afterward, as Beugnot walked home toward the Palais-Royal, he felt a flicker of sympathy for the poor cardinal, besieged by two predators who, the lawyer believed, had "joined forces to push him into the abyss."

Beugnot was misjudging the closeness of their connection; Cagliostro didn't trust Jeanne, and she fully intended to shove him into the abyss with the cardinal. The idea that Cagliostro might serve as an innocent fall guy for the swindle grew in her mind as she noticed that he was a magnet for gossip. Many of the wilder stories about him were actually recycled legends borrowed from earlier magicians—that Cagliostro's elixir had accidentally turned a chambermaid into a baby; that he was the legendary Wandering Jew who'd traveled the earth for 1,400 years; that he had chatted with Christ on a beach in Galilee; that he could make diamonds and jewels grow to twice their natural size. The man was obviously too fantastic to be completely respectable; it wouldn't be too hard to suggest that only a phony could boast such marvels.

Masonic socialites might gush over Cagliostro, but Jeanne knew that the police would always regard him as a crook. Simeon Hardy, a patron of gutter journalists, noted rumors in August 1785 that Cagliostro was both a swindler and a spy. Seraphina, too, was widely thought to be Rohan's new mistress. Hack writers featured Cagliostro and Seraphina as orchestrators of Masonic orgies. One pamphlet—attributed to Cagliostro's valet—presented the higher degrees of Egyptian Freemasonry as a series of Herculean sexual ordeals. Ordeal number three, for example, had Cagliostro watching while Seraphina made love to a seven-foot giant with a monster erection. Later in the same tract Cagliostro was lowered naked from the ceiling of the Isis Lodge, clasping in his hand "a snake of honest size and good length." So armed, he urged female Masons to return to nature by removing their clothes and beginning a collective orgy.

Jeanne also believed she could avert possible exposure of her role in the swindle by getting the compromised cardinal to pay for the diamond necklace himself rather than allow a damaging scandal. The mere presumption of writing love letters to the queen of France—even a bogus queen—was enough to get him executed. Much better to pay up and shut up. And if this strategy happened to fail, she could always blame the dodgy magician.

As a result, she showed no particular concern when the deadline—1 August—for the queen's supposed payment of the first installment to the jewelers drew close. Instead of getting out of the country, as her husband Nicolas and her lover Rétaux were soon to do, Jeanne chose to go on a spending spree. First she bought an expensive new property at her Valois hometown, Bar-sur-Aube. Then, throughout the latter part of July, nobles of Champagne watched her servants unload wagons filled with expensive furnishings, paintings, books, clothing, statuary, carriages, clocks, horses, and jewelry, including a mechanical canary worth 1,500 livres that flapped its delicate jewel-encrusted wings across the room.

Jeanne's plan nearly worked; the cardinal later admitted that he would certainly have paid for the necklace to cover up the scandal. However, the jewelers panicked because the deadline had passed, ignored Jeanne's advice, and approached the queen directly to ask for payment.

Jeanne was at a dinner party with the abbot of Clairvaux in Bar-sur-Aube on the evening of 17 August, when a latecomer dropped a bombshell among the guests. Have you heard? Cardinal Prince Rohan has just been summarily arrested in Paris—something to do with buying and stealing a diamond necklace in the queen's name. Shaken, Jeanne headed home, asking Beugnot to show his devotion again by helping her to burn bundles of incriminating letters, most of them from the cardinal to the queen. Some of the contents were too sizzling even for Beugnot's strong stomach. Amid the smell of burned paper, perfume, and molten wax, he begged her to leave France immediately, but she coolly assured him that there was nothing to fear. The whole affair was the work of that bizarre sorcerer Cagliostro.

She repeated this refrain on 20 August, after she too was arrested, taken by coach to the Bastille, and lodged in one of the cells. In a tone of bewildered innocence, she told the police that she knew nothing of Queen Marie-Antoinette or a fabulous diamond necklace—as a penniless Valois orphan, how could she know anything? Any swindle must certainly have been perpetrated by the charlatan Cagliostro: his hold over the cardinal was notorious; his wife was the cardinal's mistress; and he lived in impossible opulence. What could be more suspicious than a man "who never having gathered anything, or sold anything, nonetheless...owns everything"?

The police agreed. Three days later, Police Commissioner Chesnon and eight constables seized Cagliostro at his house in rue St.-Claude and marched him ignominiously down the boulevard St.-Antoine to the Bastille. He was placed in a cell called La Calotte within tower block D, the Tower of the Comte. He didn't know that his darling Seraphina was arrested a few hours later and locked up in a different tower.

In the weeks and months that followed, Jeanne refined and adapted her stories with virtuosic brilliance. From the outset she adopted the dual roles of victimized orphan and imperious princess of the blood. As a Valois, she insisted on a silver plate to eat off and a feather bed to sleep on. She'd had the foresight to sneak off a message to "Baroness d'Oliva," ordering her to flee the country so as not to risk her life. Poor bewildered Nicole set off for Brussels immediately. Nicolas La Motte and Rétaux had already left. Now that Jeanne's accomplices were safely out of the way, she concentrated on blaming the Copt and the cardinal.

The many and shifting stories that she peddled to police interrogators always contained one particular vignette, a rococo version of the séance with her niece. What a gothic scene she painted. The virginal girl had been mesmerized in the cardinal's candlelit boudoir by means of a black-magic ritual involving Arabic chants, whirling swords, Masonic ribbons, cabalistic fetishes, foot stompings, and secret oils. Cagliostro's diabolical performance, she said, had aroused the cardinal to gibbering ecstasy. Both men had winked and leered as though intimate with the queen. Proof of this association had come at the end of the séance: Rohan and the sorcerer handed her husband Nicolas La Motte a whitewood box filled with diamonds which they instructed him to sell in Britain.

Jeanne's second strategy took shape within a legal memoir drawn up and published by her enamored defense counsel, Maître Doillot. Its chief aim was to demolish Cagliostro's credibility as Mason, healer, and alchemist. Like a magician herself, Jeanne made as if to whip off the Copt's mask and reveal a swindler lurking behind it. Who did this man claim to be? He said he was a supernatural being alive since the feast of Cana, "initiated in the cabalistic arts, especially in those which astound ordinary people by seeming to communicate with the dead and the absent...one of the extravagant Rosicrucians who know all the human sciences, are expert in the transmutation of metals, and principally of gold...and someone who supposedly treats the poor for nothing and sells immortality to the rich." But what was he really? Nothing but a charlatan of dubious origins; a Portuguese Jew, Alexandrian Egyptian, or Greek adventurer; wanted for crimes all over Europe. Even his fantastic claims had been stolen from earlier magicians.

Jeanne's next published memoir presented Rohan as the originator of the plot but inflicted fresh damage on Cagliostro by incorporating the inventions of the embittered druggist Carlo Saachi in Strasbourg. She described Cagliostro as a former valet de chambre from the slums of Naples whose real name was Thiscio or Ticho. A syphilitic barber's son who knew nothing of medicine or magic, he'd entirely concocted his exotic origins. Egyptian Freemasonry was nothing more than a cover for swindles and obscene practices that profaned the holy Catholic church. Furthermore, the cardinal had heaped gifts of jewelry on Seraphina; obviously he'd stolen the necklace to please this latest mistress.

Jeanne's second legal defense became a best-seller in Paris; Doillot's house had to be protected by troops to stop the mobs from bursting in to get copies. Five thousand were sold in the first week, and 3,000 citizens wrote begging to buy the memoir. Worryingly for Cagliostro, Jeanne was also being helped by the most powerful political faction in the country, led by the chief minister, Baron de Breteuil, on behalf of the queen. Both Breteuil and

Marie-Antoinette loathed Rohan, and they were determined, as Jeanne later said, "to saddle him with everything." If they could convict the cardinal's guru in the process, so much the better. Abbé Georgel found himself blocked from tracking down Jeanne's accomplices because, he claimed, Marie-Antoinette didn't want either the cardinal or Cagliostro to be exonerated. Matters looked bleak.

Yet Jeanne had taken on a tough opponent; choosing to duel with Cagliostro using the weapons of storytelling and performance was a game that played to his strengths. From the moment of his arrest until his acquittal nine months later, Cagliostro fought on high moral ground. He presented himself as a man of devout spirituality and benevolence—and there were plenty of people to attest to it. This was no Sicilian tough but a man of feeling who burst into painful sobs as soon as he was taken to his cell. So damaged were his nerves by the shock of incarceration that someone had to stay with him night and day in case he tried to harm himself. (This also gave him a companion with whom to chat and play cards.)

His intercepted letters to Seraphina were utterly heartfelt. On 14 November he wrote home to his "darling wife," not realizing she was in the neighboring tower, warning her that he was mortally ill and missed her beyond life itself: "Good-bye from your unfortunate husband who dies adoring you." He did mention in passing that the pâté, partridges, chestnuts, and oranges given to him by disciples had brought partial consolation, though he was pining for his favorite meal of macaroni.

Cagliostro elaborated this benevolent identity during two formal police interrogations in August 1785 and January 1786. His answers blended edited facts with a little judicious invention. He was aged around thirty-nine, having been born, he thought, to noble parents in Medina who'd died when he was a baby. He'd qualified as a doctor by studying botany and chemistry for eighteen years, after which he traveled and taught in Turkey, Asia, and Africa before eventually landing in Rome. There, he'd married Seraphina, befriended Cardinal Orsini and the pope, and distributed medicines free to the poor—a pattern of philanthropy he had continued in Saint Petersburg under the protection of Potemkin, and in Strasbourg with the support of Marshal Contades. He even tendered a letter of recommendation from the present foreign minister of France, Comte de Vergennes, describing Cagliostro as someone famous for his benevolence in Strasbourg.

Nowhere did he mention the touchy subjects of Freemasonry and alchemy. He portrayed himself as a devout Roman Catholic; his mission had always been to manufacture medicines and elixirs that would reduce pain, heal the ill, and extend human life. Jeanne's trump card, the diabolical séance, he persuasively dismissed as a harmless experiment in mesmerism, a piece of parlor play that had nothing whatsoever to do with diamonds. As for the mysterious sources of his wealth, he reported the truth, that bankers in Switzerland and Lyons had given him generous drafts in gratitude for his healing services.

Taken by the police interrogators to confront Jeanne in person, Cagliostro unleashed towering breakers of rhetoric, eventually driving her to such fury that she grabbed a candlestick and hurled it at his stomach. By Jeanne's own account, Cagliostro retaliated with a thundering prophecy: "He will come, thy Villette [Rétaux de la Villette]. He will come; it is he that will speak." This prediction that her key accomplice would blow the whistle on the swindle proved uncannily correct. Through tireless detective work, Abbé Georgel eventually tracked Rétaux to Geneva, where he was captured on 26 March 1786. When Rétaux in turn was faced with Cagliostro, the swaggering guardsman collapsed in a heap. Rétaux was utterly chastened by an hour-long lecture on his moral duty before God. Soon afterward, he made a new statement admitting his role in the forgeries.

Cagliostro also attracted an unexpected source of help against Jeanne. Soon after entering the Bastille, he was offered the support of an anti-Bourbon political faction led by a reformist lawyer and parliamentarian, Duval d'Épresmésnil. Behind the lawyer lurked the money and influence of two great political clans, the Rohans and the Orleanists. The latter were led by the scheming Prince Louis-Philippe-Joseph, duc d'Orléans, a patron of sedition who hated the queen for personal reasons, and his cousin the king for dynastic ones. Orléans was a man who thought he should be king.

As a member of the parlement, d'Épresmésnil had to entrust the actual preparation and writing of Cagliostro's published defense to his brother-in-law Jean Thilorier, a rising young legal star with great literary flair. If Jeanne's lawyer could turn her into Cinderella, Thilorier could make Cagliostro into Aladdin. The Copt's birth, the lawyer admitted in the first memoir he wrote for Cagliostro, was shrouded in obscurity. It seemed likely he was the son of the mufti Selahayam at Medina or of the sharif of Mecca, both of whom had loved him as their own. The boy was raised as a Muslim prince under the name of Acharat and given a guardian called Althotas to instruct him in the secrets of the pyramids hidden in "vast subterranean caves dug out by the Egyptians, to be made the repository of human knowledge."

Acharat had then spent some years in Rhodes and Malta under the tutelage of the Knights of Saint John. Here he resumed European dress, adopted the name Cagliostro, and acquired his devotion to Catholic doctrine. From that time on, he'd been seized by a compulsion to help suffering humanity: "I found means to preserve my independence by giving, but never receiving; my delicacy was such that I refused rewards, even from the hands of Sovereigns,—the wealthy I have assisted with remedies and advice—to the poor I have given both money and remedies.... A stranger everywhere, I have everywhere fulfilled the duties of a good citizen—everywhere I have respected the religion, the laws and the government.—such is the history of my life."

Over the previous twenty years, reformist French lawyers had been using strategic cases as theaters of political contest and entertainment, in which the public rather than France's governing elite gradually became the real judge. Though the Bourbon monarchy had not yet realized this, the best-selling genre of legal memoirs was already eroding its absolute and divine authority. Thilorier was said to have laughed in private at the absurdities of Acharat: he shared Beugnot's opinion that publishing half-fictive legal briefs had turned France's legal system into "a circus." Bawling and bullying, mawkish and savage, compassionate and cruel, public opinion had thrust its way into the old regime.

The diamond necklace affair was, of course, the greatest circus of all. Each day newspapers all over Europe carried steamy details of its troupe in the Bastille, interleaved with speculations about the puppeteers behind the scenes. In Saint Petersburg, Catherine the Great pored over newspaper accounts of Cagliostro's memoir and congratulated herself on having driven the dangerous shaman out of Russia. In Mitau, Elisa von der Recke began to assemble her journal notes to tell the world of her encounter with this famous necromancer. In Dux, Bohemia, raddled old Count Casanova began writing his own exposé, envious of the attention being heaped on the Sicilian adventurer.

Though the daily papers in Paris were still controlled by royal censorship, an illegal underground of printed gossip, innuendo, pornography, satire, cartoons, and song had grown up to feed popular tastes. In shabby little courts like the rue des Fossés-St-Bernard, men like the barber's assistant Louis Dupré and the bookseller's agent Antoine Chambon churned out cheap works filled with tales of the lurid entanglements between Copts, cardinals, queens and whores.

This transformation of the French judicial process into a form of popular spectacle perfectly suited Cagliostro's talents as a showman. Once he realized that he was relatively safe from conviction—he'd arrived in Paris from Lyons after the necklace was already sold—Cagliostro began to relish the media's attention. Thilorier's exotic *Memoir for Count Cagliostro* sold as briskly as Jeanne's. Once again, soldiers had to be posted to control the crush of people trying to buy copies.

Seraphina published a comparable defense. According to her lawyer Maître Polverit, Countess Cagliostro was "an angel free of sin," "a model of tenderness, sweetness and resignation," and "an ideal of perfection." What was such a being doing in the terrible Bastille alongside grubby courtesans like Jeanne La Motte?

Cagliostro asked the same question in a celebrated letter of 24 February 1786. Addressing the parlement directly, he pleaded for Seraphina's release because she was gravely ill: "Will you let her perish when she could receive the aid of a beneficial art practiced by her husband?... Will you condemn her to perish alongside her unfortunate husband, without [his] being able to give her care or consolation?" He omitted to mention that Seraphina was actually living in rude good health in the Liberty tower, enjoying the personal attentions of the prison governor and of her maid Françoise. Parlement ordered her release on 18 March, whereupon Seraphina found herself a sensation in society.

As the interrogatory evidence mounted against Jeanne, her performances in the Bastille became more extreme. She threw violent fits, bit the prison warder on the hand, tore off her clothes and lay naked on the bed refusing to move and shrieking abuse against all those who dared testify against her. Over the course of numerous interviews, she twisted, feinted, and doubled back on herself, always looking for new targets to accuse. At one time or another she pinned blame on almost every figure caught up in the scandal, including her sister, her husband, and her lovers. Her favorite candidates, however, remained the Copt and the cardinal.

In particularly desperate moments Jeanne began to grope toward a last awesome accusation. She hinted that any contradictions in her testimony stemmed from a need to protect the reputation of someone *very eminent indeed*. With a possible charge of treason hanging over her head, she didn't dare to say the name of Queen Marie-Antoinette directly, but by the time of the actual trial on 31 May, she was prepared to tiptoe down this path.

Sitting upright in the criminal chair (known as *la sellette*) before the full parlement of Paris, wearing a dress of lavender satin with black velvet trim, Jeanne put on the performance of her life. With an air of deep sincerity she testified that Rohan had shown her more than 200 intimate letters between himself and the queen, and that he had several times met the queen alone at night in the Trianon Palace. At least once, Jeanne ostentatiously refused to answer a question because it "might offend the queen." With her acute gutter instincts, she sensed that Bourbons' enemies wanted to hear a story in which the queen and the courtesan had become interchangeable.

None of the four people who waited with such terrible anxiety on the evening of 31 May to hear the verdict in the trial of the diamond necklace affair ended up victors in the long term. So distressed was Marie-Antoinette at the cardinal's acquittal that the king tried to appease her by agreeing to acts of petty vengeance. He ordered Cardinal Rohan to relinquish all royal titles and simultaneously issued a *lettre de cachet* banishing him permanently to the Abbey of the Chaise-Dieu. Here, in the remote Auvergne mountains, hemmed in by snow for much of the year and restricted to the company of forty Benedictine monks, he would have no chance of enjoying the sybaritic pleasures of the past.

Count Cagliostro was given only a day to savor his popularity before a messenger from the king arrived at the door of 1 rue St-Claude with a decree of banishment. He was told to leave his lovely house in Paris within three days and leave the kingdom in three weeks. The couple moved immediately to the château of the marquis de Boulainvilliers in Passy—ironically, this was where Jeanne had begun her career as an adventuress. It was now a headquarters of mystical Freemasonry. They were followed all the way by throngs of Parisian supporters on foot and in wagons. A visitor from Bâle, who watched the procession, recorded that street vendors could not keep up with the demand for souvenirs of Cagliostro. As well as the usual snuffboxes and portraits, people eagerly paid for fans like those Balsamo had once sold—decorated now with scenes of the Great Copt tending to the poor and stoically enduring the rigors of the Bastille.

On 16 June 1786, the Cagliostros left Boulogne-sur-Mer on a boat bound for Dover. Cagliostro's later memoir tells the story: "The shores that I quitted were lined by a crowd of citizens of all classes, who blessed and thanked me for the good I had done their brethren, addressing to me the most touching of farewells. The winds carried me far away from them, and I heard them no more."

Jeanne's sentence was carried out early in the morning of 21 June. She was woken so suddenly that she had time only to put on a petticoat and shawl before she found herself seized by eight men, bound, and taken before the registrar for her sentence to be read out. A blow on the back of the legs and a thump on the top of the head brought her to her knees. She was dragged to the courtyard of the Palais de Justice, which had been set up with benches for the public spectacle. Because it was so early in the morning, however, only a few hundred people had gathered to watch. Never had the Parisian executioner faced such a formidable victim. Jeanne fought like the famished tiger she was. After she had been tied to a cart with halter rope around her neck and whipped lightly across the neck and shoulders, she turned to face the executioner and shrieked as she saw him holding a red-hot branding iron.

The scene that followed was more like a rape than a sentence of law. As she struggled, the executioner and his assistants ripped the petticoat from her body, pushed her to the ground and thrust the scorching iron into her shoulder. Lying nearly naked before the prurient gaze of the crowd, she twisted her body just as the iron descended the second time so that it scorched a V on her breast. Before passing out, she managed to bite a chunk out of the executioner. Bystanders claimed that they heard her scream obscenities against Cagliostro, the cardinal, and the queen.

Half-unconscious, she was bundled into a carriage for transportation to the infamous Salpêtrière women's prison, only to lunge suddenly toward the door in an effort to throw herself under the carriage wheels. As the nuns at the prison bathed her wounds and tried to put drops of water into her swollen mouth, she dimly heard the attendant praising her gold pendant earrings. Instantly she began bargaining with him—even in her stricken state, Jeanne could smell a sucker.

With the sale of her earrings at an inflated price went the last vestige of Jeanne's glamor. Her glossy chestnut hair was shaved off, and she was forced into the standard prisoner's dress of coarse gray drugget, rough petticoat, penitential cap, and wooden sabots. Armed guards then led her through the first two divisions of the prison—allocated to prostitutes and unwanted women respectively—until, after passing nine locked courtyards, they reached the section reserved for female lifers. There for the first time she saw the dormitory where she was expected to see out the remainder of her days. Tough as she was, Jeanne flinched at the sight and sound of the troop of scarecrows who rushed up to inspect the latest inmate. The first lesson they were eager to teach her at this "seminary of vice and depravity" was the infamous perversion known locally as the "sin of the Salpêtrière."

In the years to come she would openly accuse Marie-Antoinette of the same sin. This was one of the malicious charges that contributed to the queen's demise. It was thus Marie-Antoinette herself who ultimately paid the highest price of this tawdry affair: she lost her dignity, a diamond necklace—and, eventually, her head.

· 5 ·

Prophet

Prophesy: To speak by (or as by) divine inspiration, or, in the name of a deity, to speak as a prophet; to utter predictions, to foretell future events (by inspiration or generally).

EARLY IN THE SUMMER OF 1786 British newspapers began reporting a rumor that Count Cagliostro was about to arrive on their shores. Most were welcoming; they cited his fabulous achievements, exonerated him of involvement in the affair of the necklace, and wondered excitedly about his intentions in Britain. The *Times*, typically, praised the count as "an excellent scholar" who practiced medicine, "not with a view to gain, but merely upon a principle of humanity and goodwill to his fellow creatures."

The climate had been warmed beforehand by an English translation of the count's famous trial brief, *Memorial...for the Comte de Cagliostro* (1786), which described his exotic Arabian origins and rebutted Jeanne La Motte's vicious charges. The *Times* thought this memoir "perhaps the most extraordinary publication that ever engaged a body of people. It is read by the lowest mechanic." Its British editor and translator, Parkyns Macmahon, a defrocked Irish priest, tantalized his readers by claiming to settle the many speculations about the Great Copt's fabulous origins. Cagliostro had actually been born, he authoritatively pronounced, to a noble family living within the ancient city of Trebizond on the Black Sea in the Turkish Empire.

Macmahon's source for this discovery was a newspaper, the *Courier de l'Europe*, which he described as "the only criterion of authenticity" for French news. As a former subeditor of the paper, Macmahon was biased, but the *Courier* was certainly a staple for anyone interested in Anglo-French affairs. Published twice weekly from London in both English and French editions, it sold 5,000 issues per week. In the eighteenth century this was a large circulation; and by the time the paper passed through a circuit of coffeehouses, taverns, and salons on both sides of the Channel, it had reached at least four times that number of readers. Brissot de Warville, a French journalist who'd also worked on the paper (and would become famous during the revolution), put its readership at more than a million. Even if he was exaggerating, the *Courier* hit precisely the audience Cagliostro hoped to influence.

Though Cagliostro expected to find fewer disciples in London than in Paris, he hoped that Britain would greet him with something like the enthusiasm he'd attracted elsewhere. After all, this time he was coming not as Giuseppe Balsamo the shady Italian painter or as Colonel Pellegrini the bogus Brandenburg soldier, but as Count Alessandro di Cagliostro—a legendary Oriental healer and scientist, martyr of French tyranny and international Masonic star.

The welcome exceeded his expectations. As soon as he reached London, around 20 June 1786, Cagliostro found himself feted by the most glittering set of Freemasons in the country. Britain's famously unfilial royal sons—George, Prince of Wales; Edward, duke of Kent; and William, duke of Clarence—were among the earliest guests to show their Masonic insignia. All belonged to distinguished lodges in London. Freemasonry was a favorite position from which to snipe at their father, King George III.

The Prince of Wales, the most senior of the brothers and heir to the British throne, contacted Cagliostro at the urging of a scheming royal prince, the duc d'Orléans, from the other side of the Channel. Orléans had visited London a month or so earlier on the trail of the diamond necklace: he was hoping to dig up some fresh dirt to use against Queen Marie-Antoinette and her husband. On 5 April 1786, while questioning a jeweler in Bond Street who'd bought most of the stolen diamonds, Orléans happened to bump into his fat English friend, Prince George. They set off immediately on a royal binge.

Along with a shared love of fast living and political intrigue, the two princes were fascinated by Freemasonry. Orléans, as head of the Grand Orient of Paris, had not long before inducted the Prince of Wales into his lodge—at around the same time that Cagliostro had initiated Orléans and his mistress into the Egyptian Rite. Now, on his friend's advice, George sent a personal Masonic mentor to make contact with the mighty Copt.

British Masonic dreamers had also looked forward to the arrival of the Great Copt. General Charles Rainsford, an initiate of no fewer than eleven mystical lodges, had long wanted to make contact with the famous seer. The duke of Northumberland, a fellow visionary, had written the previous year to tell Rainsford that "...the famous Count Cagliostro is in Paris, and... the manner in which he lives, and the stories told of him are so wonderful that I long much to know whether there is any truth in them.... He is said to be 300 years old, and lives without any visible means of acquiring such wealth, at greater expense than the first of the Nobility of Paris." Rainsford was among a handful of British delegates who attended the mystical summit in Paris that had so fatefully attracted Cagliostro from Lyons at the end of January 1785. Like most of the Masons present, Rainsford had been disappointed that the meeting failed to reach some accommodation with the Egyptian Rite.

British occultists like Rainsford frequented a variety of small mystically-minded lodges in central London where clusters of artisans and intellectuals gathered to dream, argue, and experiment. During the 1770s many of these seekers had been inspired by the cabalistic teachings of a mysterious turbaned alchemist from Poland, Dr. Samuel Jacob Falk, known in mystical circles as the "B'aal Shem of Wellclose Square." Others had been influenced by the famous celestial writings of the immigrant Swedish engineer Emanuel Swedenborg. Some, too, had been introduced to theosophical ideas by visiting Illuminés from hubs of European mysticism in Avignon, Paris, Stockholm, and Berlin.

It was this milieu in London that nurtured the genius of the brilliant engraver-poet William Blake and his wife, Catherine, then living on the south side of the Thames at Lambeth. Foreign-born artists like Philippe de Loutherbourg, Lambert de Lintot, and Francesco Bartolozzi were also drawn to the heady experimental atmosphere of these gatherings. Whatever their Masonic allegiance, most of London's mystics longed to penetrate the veil of the spirit world by means of vision, dream, or trance and rediscover the ancient sources of magical healing and alchemical transmutation.

Cagliostro's friends and admirers expected him to set up a healing center in London along the lines of his clinics in Russia, Alsace, and France. He was pleased to find some local sponsors willing to help. Thanks to an introduction from a senior Egyptian Mason in Paris, the Cagliostros were overwhelmed by the hospitality of Samuel Swinton, one of London's leading press proprietors. A Scot from Ilk of distinguished family, Swinton had spent his boyhood in the navy and then lived for some years in France before moving to London to become a considerable businessman. By 1786, he was a proprietor of the *Courier de l'Europe* and had major interests in several other British dailies. He also ran a flourishing wine business, a boardinghouse for young immigrants, a French restaurant, and a nursery for foreign plants cultivated in his substantial gardens in Knightsbridge.

The friendly Scot took the Cagliostros on excursions to see the city's most picturesque sites, including the sublime sweep of the Thames from the hillside near Greenwich Hospital, a view much admired by landscape painters. Better still, he gave the Cagliostros a vista of their own by renting them a large villa with rolling lawns and a lovely garden. The address was 4 Sloane Street, Knightsbridge, next door to Swinton's own house. His

French-born wife Felicité quickly befriended Seraphina and volunteered their five children to help the visitors move in. Nowhere had the Cagliostros received kinder treatment.

Although the new house needed some refurbishing to bring it up to Seraphina's standards, she loved spending money; and Swinton was prepared to help her obtain fine imported goods through his many business interests. Before long, she'd made the house look—as one German visitor said—like an "enchanted castle." The interior had luxurious carpets from Persia, mahogany furniture from India, porcelain dinnerware from Italy, and a cellar stocked with costly French wines. She hired an Italian chef named Agostino to prepare macaroni the way Cagliostro liked it, a black valet in Moorish costume to announce their guests, and a chambermaid and a hairdresser to look after her own needs.

On the purely business side, Swinton hoped that Cagliostro would eventually join him in opening a major clinic like the one in Strasbourg. Before rushing into such a large investment, however, he suggested that they begin with a smaller-scale partnership. Swinton would hire a reliable apothecary to make up and sell the Copt's red Egyptian pills, excellent for rejuvenating both body and soul.

A few days after the Cagliostros moved in, Swinton strolled over to introduce them to a friend who was especially eager to meet them. He was Théveneau de Morande, a sharp-eyed Frenchman from Duke Street who'd worked in London for many years. As editor of the *Courier de l'Europe*, de Morande had joined in the general speculation about Cagliostro's visit to Britain. "This extraordinary man has come," he wrote on 23 June 1786, "to exercise his talents in London. Only a great theatre seems to suit his activities; it is evidently some vast scheme that has brought him here." Anyone who knew the journalist would have noticed his ambiguous use of the words "vast scheme." Did he mean a grand plan or a grand swindle?

Cagliostro wasn't sure what to make of the man. De Morande was silkily charming to Seraphina but disturbingly insistent on questioning them

Théveneau de Morande

about their previous visits to London. Clearly, one would have to watch this character.

Cagliostro was wise to be wary: Théveneau de Morande was not just any journalist. He was Europe's most dangerous pen for hire. No modern tabloid writer could surpass him for savagery. French contemporaries called him *le Rousseau de ruisseau*, the gutter Rousseau, and often worse.

A lawyer's son born at Arnay-le-Duc in Burgundy in 1741 and christened Charles Théveneau, he was almost the same age as Cagliostro and equally well-drilled in roguery. He'd learned something of the law before being arrested for stealing from a brothel. After that, he'd spent some time in the army, then drifted into the shadows of the Parisian underworld, where he'd worked as a pimp, blackmailer, and con man. Handsome, quick-witted and articulate, he'd attached himself to a succession of prosperous courtesans and easygoing society women. This led to a stay in the Bastille, followed by several years of compulsory internment in a convent in Armentières. Police Inspector Marais, in a letter to a colleague of 5 May 1768, made an additional accusation: "Théveneau...was strongly suspected of being implicated in the crime of homosexuality and to have worked as a tout for these villains."

The following year, around the time Giuseppe Balsamo was scooting from Palermo just ahead of the police, twenty-four-year-old Charles Théveneau was making a similarly motivated flight across the Channel to join London's émigré underground. By the middle of the eighteenth century the city had become a haven for French criminals escaping from their country's extradition and libel laws. The outlaws within this urban badlands included bankers fleeing from swindled clients, army officers absconding with regimental funds, and priests evading sexual charges. Most prolific of all, however, were writers ducking France's book licensing and censorship laws. A Genevan bookseller named Boissière and a disgraced French politician called Goezman (alias Baron Thurn) became supremos of the London trade, running a large stable of hack writers, printers, and publishers. They churned out libelous pamphlets and "scandalous chronicles" in French, English, and Dutch, an output with the double aim of being sold in Europe's book markets and being used to extort suppression fees from those they libeled.

Here Charles Théveneau had found his natural habitat: he invented for himself the noble lineage of "de Morande" and perfected the art of writing viciously pornographic tracts called *libelles*. He married his London landlady's daughter, Elisabeth, and produced a succession of children—he told an early blackmail victim, the transvestite soldier-diplomat Chevalier d'Eon, that he worked not for wealth, fame, or glory but to sustain a sickly wife and hungry children.

Soon, though, he'd begun to stand out from the Grub Street pack (which included Parkyns Macmahon, who would become Cagliostro's English biographer). De Morande gave his first London publication the apt title *Le philosophe cynique* (1771); its success spawned a more ambitious work called *Le gazetier cuirassé* that became a byword for nastiness throughout the francophone world. De Morande's wit, savagery, and wide network of informants made the journal unusually lethal. A contemporary English print showed a victim decapitated by the editor's pen while nests of vipers writhed among the pages. Brissot de Warville, a former hack himself, neatly described its mode of operation: "The victim pays to have his name kept out, his enemy pays to get it put in again; the law protects everything and the public does nothing but laugh."

A few of de Morande's early victims of blackmail didn't actually play by these rules. Théveneau earned a literary thrashing from Voltaire, a horse-whipping from the count de Laureguais, and a challenge from the deadly swordsman Chevalier d'Eon. But no one could deny his stamina: he eventually hit a vein of gold in 1774 by threatening to publish a pornographic biography of Louis XV's mistress Madame du Barry (for whom the infamous diamond necklace was originally commissioned). At first, Louis tried to silence this gadfly by sending a gang of undercover French police to London to kidnap him, but de Morande incited the patriotic local crowd to heave the policemen into the Thames. Defeated, Louis sent over a gentler negotiator, the writer-playwright Pierre-Augustin Caron de Beaumarchais. Posing under the name Ronac, Beaumarchais devised a settlement that gave de Morande the huge sum of 32,000 livres (and a lifetime pension of 4,000 livres per year), provided that he withdraw the publication and stop writing against the French crown.

This didn't deter de Morande from smearing other dignitaries; his *libelles* made him one of France's five best-selling writers during the last years of the old regime. In 1784 he became editor of the *Courier de l'Europe* and his poisonous tentacles began to extend to other parts of Europe as well. Within Britain, his skill at flattering or blackmailing the middle echelons of both the government and the opposition made him one of the best-informed men in politics. His journalist's snout also took him deep into London's pleasure world, truffling for usable filth among brothels, taverns, casinos, and racecourses.

As is often the way, the crook turned policeman too. When de Morande took over the editorship of the *Courier*, the paper began to collect a subsidy from the French government. The foreign minister, Comte de Vergennes, boasted that "the *Courier* is worth a hundred spies to me and costs a good deal less." Three years, earlier, de Morande had become a spy in a more technical sense. He joined the Bourbon secret service in place of a man whom the British hanged for espionage during the American war. Whether de Morande had sold out his predecessor remains unclear, but there's no doubt that he took to spying like a master. Over the next ten years he was paid handsomely for sending the French government regular packets of intelligence. His detailed reports covered British commercial inventions, naval and military deployments, and the activities of the French immigrant community in London, particularly those involved in the cross-Channel smut trade.

Count Cagliostro knew none of this, and like many truly devious people, he could be remarkably naive. As on earlier visits and despite the warm welcome, the Copt continued to have difficulties with the British. Their language, manners, and values baffled him. He couldn't understand the country's thrusting commercial ethos, its snobbish yet oddly fluid social structure, its aggressive Puritan traditions, or its kaleidoscope of political factions. Intoxicated by his own mystique, Cagliostro accepted his new batch of London admirers trustingly, equating them with the adoring disciples who'd flocked to his clinics in Strasbourg, Lyons, and Paris.

Cagliostro also failed to realize, until too late, that his recent imprisonment in the Bastille had changed his public image in a fundamental way. It was not just that he'd become an international star, part of a spectacle-hungry celebrity world; this he could handle. But the necklace affair had caught him in a tide of popular politics that would eventually sweep away France's old regime. When the radical lawyers Duval d'Eprésménil and Jean Thilorier had offered to defend Cagliostro in the Bastille, he'd naturally jumped at the chance. Though sensing vaguely that the lawyers might have motives of their own, Cagliostro failed to see that they'd grafted him onto a political campaign to hobble the powers of the king and reform the parlement. D'Eprésménil called it "de-Bourbonizing." Radical lawyers were especially keen to cauterize the king's power of imprisoning people by arbitrary *lettres de cachet*; and Cagliostro's sensational imprisonment made him a perfect weapon for dramatizing French monarchical tyranny. Without realizing it, Cagliostro had permitted himself to become a creature of politics, and there was a price to pay. From now on he would always be viewed through a lens of revolution or reaction.

Cagliostro assumed that his acquittal would allow him to slip back into his glittering former life, but his lawyers wanted their reward. The day after his release, they persuaded him to pose with another recent celebrity, Passy la Salmon, a poor peasant girl from Rouen who'd just escaped the death sentence through the intervention of the parlement. D'Eprésmésnil and Thilorier now urged Cagliostro to bring a case for damages against the governor of the Bastille, de Launay, and the police chief, Chesnon. Hadn't they stolen his precious elixirs and recipes? Yes, of course they had, and Cagliostro could do with some financial compensation. The lawyers, how-ever, were more interested in landing another blow on the already ailing Bourbon state by attacking two major symbols of repression, the police chief and the prison governor.

On 20 June 1786—the day that Cagliostro arrived in London and Jeanne was scorched by the hangman in Paris—the two French lawyers filed a civil case on Cagliostro's behalf at the Châtelet. Their client sought heavy damages—150,000 livres—for the deliberate or negligent loss of precious jewels, elixirs, and recipes, plus a further 50,000 livres in compensation for the rough handling he'd received on the way to prison. A pamphlet detail-ing his accusations appeared in the bookshops in Paris on the same day. At Versailles, Louis XVI was stunned by this new provocation, and the baron de Breteuil became so angry that he ground his teeth whenever he saw Cagliostro's picture in the Parisian bookstalls.

An explosive letter dated the same day and signed by Cagliostro appeared on both sides of the Channel a week later. Purportedly sent by him to a friend in Paris, it described how the great healer had been betrayed by the rulers of the nation to whom he'd given his heart and soul. Cagliostro probably didn't write the letter himself, but he dictated the gist of it to his lawyers, who then gave it a good political polishing. Whatever his later regrets, the Copt was proud at the time that the letter was written "with a freedom rather republican." Manuscript versions circulated in anti-Bourbon circles for several months until mid-1786, when it was printed in French and English under the simple title *Count Cagliostro's Letter to the French People*.

This was a fateful moment for the Copt. The letter threw Cagliostro into the maelstrom of subversive politics. He was helped by the fact that literature from the Bastille had already become a best-selling horror genre— just as the prison itself had already become a symbol of archaic despotism. Readers of the letter on both sides of the Channel expected to find descriptions of torture, isolation, and terrible hardship; and he obliged them. He told them at one point that he'd choose the death penalty rather than face another six months in that Miltonic hell.

In the middle of these gothic passages, however, he'd suddenly switched to prophecy. "Somebody has asked me whether I would return to France if the ban was lifted," the letter said rhetorically. "Certainly,... provided the Bastille was turned into a public promenade; would to God it could be done." When the Bastille fell to the revolutionaries on 14 July 1789, these words seemed eerily prescient. Other concrete predictions also seemed to come true: "It is a fitting purpose for your parliament to work [for] the necessity of revolution.... This revolution, so much needed, will be brought about, I prophesy it for you. A prince will reign who will gain glory by abolishing *lettres de cachet*, by convening the States-General and, above all, by restoring the true religion. This prince, dear to Heaven, will realize that the abuse of power, in the long run, destroys power itself."

After the event, Cagliostro's readers forgot that he'd been trotting out a pretty standard reformist agenda of the time. They didn't realize that the word "revolution" meant something much milder to him than the wholesale transformation that actually occurred during the 1790s. Even the reforming prince he mentioned was a reference not to Louis XVI (who introduced the reforms) but to Louis' shifty Masonic cousin, Orléans, whom Cagliostro was rather naively backing. But none of this mattered. As each day passed, events in Paris seemed to click into line with predictions in the letter, making Cagliostro look like a master of the future.

The letter also contained some other grenades for his enemies. The chief minister, Baron de Breteuil, took a pasting. Cagliostro called him a vengeful persecutor who'd slimily covered up the real truth of the diamond necklace affair; and he promised to bring the man to justice along with his fellow conspirators, de Launay and Chesnon.

The letter hit its mark. Chesnon said that it began France's tumble into revolution. Breteuil was so troubled by the contents that he dragged his gouty frame to Versailles to advise the king to get Cagliostro back to Paris. Perhaps he hoped this would disarm the man's criticism, or perhaps he simply wanted to lure Cagliostro back into the Bastille. Anyway, on 4 July 1786, the French minister wrote to Count d'Adhémar, the ambassador in London, ordering that the exiled Cagliostro be given leave to return to Paris while his case was being heard.

In London, meanwhile, Cagliostro had attracted a genuinely subversive prophet. Not long after the Cagliostros moved into Sloane Street, Swinton introduced them to one of his Scottish friends—a tall, pale young man with a wispy beard and a flow of words that matched Cagliostro's. Despite his sensitive appearance and noble lineage—he was a younger son of the duke of Atholl—Lord George Gordon was pure trouble. Cagliostro knew of him, naturally; everybody did. But the count did not realize that the man was ridiculed in Britain as "mad Lord George," "Lord George Flame," and "Lord George Riot." These incendiary nicknames referred to Gordon's infamous role in triggering a terrible outbreak of anti-Catholic riots in 1780.

During six days, from 2 to 10 June, the Gordon Riots had inflicted more damage on London than Paris would witness in a decade of revolution. When the historian Edward Gibbon saw the blazing prisons, looted houses, and black-coated marchers, he thought he'd been whisked back to the "dark diabolical fanaticism" of the seventeenth century. "Forty thousand Puritans such as they might be in the time of Cromwell have started out of their graves," he'd written to his sister. One of Gordon's former friends, the Irish-born political writer Edmund Burke, sensed something even more revolutionary in Gordon's rhetoric: the kilted warrior blended his anti-Catholic rant with a new language of natural rights, democratic liberties, and Scottish nationalism. Something of a prophet himself, Burke believed that Gordon represented a new and terrible future for the world. Britain, it seemed, had given birth to the first modern man of terror.

Cagliostro was oblivious of all this; Gordon's title and Masonic credentials were what impressed him. He didn't realize that the young lord's behavior was growing stranger every year. Narrowly acquitted of treason after the riots, Gordon had taken up a succession of subversive popular causes. In the twelve months before meeting Cagliostro, he'd sparked a large-scale strike of weavers in Glasgow, raised an inflammatory petition from Newgate prisoners, and infuriated the British government by protesting against its commercial treaty with the French. People who believed Gordon insane pointed also to his recent conversion to Judaism and his outspoken support for a woman who'd tried to assassinate George III with a penknife. Not surprisingly, the British and French governments were keeping the man under close surveillance.

On the morning of 20 August, after exactly two months of exile in London, Cagliostro was surprised to receive a message from the French government granting him permission to return to Paris for his impending case. If he presented himself at the French embassy the next day at eleven, he'd be given an official letter of permission from the hands of M. François Barthelemy, chargé d'affaires, standing in for the ambassador.

As luck would have it, Lord George Gordon was present when Cagliostro received the summons; he'd decided to appoint himself the Copt's protector and adviser. Shouting excitedly, he warned Cagliostro to watch out for a trap: he must not go into that lion's den alone. Gordon and a visiting French disciple would accompany him to prevent a likely kidnap attempt.

Next day M. Barthelemy was disconcerted to see the figure of Gordon, armed with a claymore, standing at Cagliostro's side. This set the tone for the meeting, which proved a disaster. After reading out the minister's letter, Barthelemy refused to hand it over because of Gordon's inflammatory presence. Cagliostro, for his part, was tense and suspicious. How, he asked, was it possible for Baron de Breteuil, a mere minister, to countermand a *lettre de cachet* signed by the king? Any warrant granting him permission to return must also be signed by the king. What proof did he have otherwise that it was not a trap? Gordon's excited interjections and waving arms added further insult. Furious, the French attaché terminated the meeting.

In spite of the paranoia of his adviser, Cagliostro's suspicions were not unreasonable. Both British and French newspapers had recently carried stories of attempts, backed by the Bourbon regime, to kidnap Nicolas La Motte. Police archives confirm that Louis XVI agreed in the summer of 1785 to send a secret expedition of French police to grab the thief by force and return him to France for interrogation and trial. La Motte was to be coaxed to an isolated house near Newcastle, overpowered in his bed, tied up in a blanket, and rowed out to a hired Tyneside collier bound for Dunkirk. In the event, the plan went awry and the Bourbons decided instead to negotiate with La Motte through Ambassador d'Adhémar.

It's unlikely the Bourbons were planning to kidnap Cagliostro, but rabid Lord George was adamant. He wrote a letter to the *Public Advertizer* of 22 August abusing the French ambassador and his staff. From now on, he declared, Cagliostro would conduct all negotiations with France through Lord Gordon. He stated that "the Queen's party is still violent against Comte de Cagliostro, the Friend of Mankind." A few days later, a second letter to a newspaper contained further provocations. Gordon asserted that the ambassador's staff were in league with "a gang of French spies," working on behalf of "the Queen's Bastille Party" to entrap Cagliostro—part of "the hateful revenge and perfidious cruelties of a tyrannical government."

By allowing Gordon to take up the cudgels, Cagliostro did himself grave damage. His friendship with a man whom respectable society shunned as a crazy incendiary immediately drained away most of his local support and spooked the British government. Newspapers had great sport lampooning the friendship between the Scottish and Arabian prophets. The Prince of Wales was one of the first of Cagliostro's supporters to disappear. George loved to goad his father by flirting with oppositional causes, but he shied away from extremism. Treason was too strong a dish for Prinny's delicate paunch. Fashionable opposition politicians like Lord Sheridan and the duchess of Devonshire, who'd earlier introduced Cagliostro into their salons, dropped him from their visiting lists.

This exodus of respectable supporters forced Cagliostro into even greater reliance on Gordon, who worsened their isolation by jealously vetting all visitors to the house in Sloane Street. One of the few people to be admitted at this time was a German writer, Sophie von La Roche, who came to London in the spring of 1786 to deliver letters from the Sarasins in Switzerland. Having earlier tried some of Cagliostro's "marvellous drops" and praised the "masculine, noble and honest tone" of his defense in the Bastille, Sophie was dying to meet the Great Copt.

She itched to discover what linked the "Asiatic" magician and the "English fanatic." Meeting them together on two occasions convinced her that they both saw themselves as divine prophets called on to renovate mankind. She also quickly discovered that Gordon's pale blue eyes and gentle manners masked an extremist's temperament. Had she been a Catholic, Cagliostro told her, Gordon would have thrown her out of the house. Paradoxically, both men attacked Catholicism in bigoted terms yet were tolerant of other religions, including Judaism and Islam. As for Lord George's Puritanism, it didn't seem much in evidence, Sophie observed, when he was flirting with Countess Cagliostro—confirming a famous catty remark once made by Lady Montagu that "the Whore of Babylon...is the only whore his Lordship dislikes."

Another bond between the two men, Sophie noticed, was a strong sense of persecution. Lord George's bony face flushed scarlet and he stuttered out a volley of insults when asked about King George III. Cagliostro seemed equally bitter about his sufferings in the Bastille—so much so that Sophie wondered if the experience had made him morbid. "If I had not that dear creature my wife," he told the novelist morosely, "I should go and live with the wild beasts of the jungle, certain of finding friends amongst them." He was particularly troubled, he told her, by the plots of "an enemy he found waiting over here in the form of a news reporter." Cagliostro had refused to pay a bribe to be puffed in the *Courier de l'Europe*, and now the paper had begun to churn out "the most futile rubbish daily."

Cagliostro might have been a good deal more depressed had he known that his journalistic enemy Théveneau de Morande was also a paid Bourbon propagandist. In late August, the French government decided to take action against Cagliostro. The lawsuit, his prophecies, and his friendship with Gordon had made him a menace. It was time to unleash the bloodhounds. Two trusted military envoys were sent to London to brief de Morande on tactics and to deliver a substantial payment. Soon, he received explicit orders from the French attaché to silence the dangerous prophet by any means possible.

Cagliostro didn't guess at first that Swinton was part of the plot. He'd begun to suspect his ingratiating neighbor of being in league with de Morande, but he knew nothing of Swinton's close connections with the French embassy. When the *Courier* attacked Cagliostro on 22 August 1786, calling him an imposter and charlatan, he still believed that de Morande was trying to extort a bribe. But if Cagliostro tried to offer one, it failed. On 1 September, the paper carried an alarmingly detailed exposé, "The Travels of Count Cagliostro Before He Arrived in France." Further installments were promised.

Having lifted most of its claims from the legal defenses of Jeanne La Motte, de Morande's article offered little new information. Nevertheless, Cagliostro was furious, and felt bound to retaliate because of the paper's influential readership. And he was not without resources. Gordon got him a spot in the *Public Advertizer*; and Thilorier, over from Paris, offered to provide any literary touches necessary.

Typically, Cagliostro chose to fight on unconventional terrain. One of de Morande's early installments had ridiculed the Copt's claim to be able to kill lions by baiting them with live animals reared on arsenic. Convinced that his enemy was gutless, Cagliostro decided to use the same tactics that had crushed Empress Catherine's physician five years before. He challenged de Morande to a special kind of duel:

> I am going, mister jester, to make a joke at your expense. In matters of physics and chemistry, arguments prove very little, persiflage proves nothing; experiment is everything. Allow me, then, to propose a little experiment that will entertain the public, at either your expense or at mine. I invite you to eat with me on November 9, at 9 o'clock in the morning. You will provide the wine and all the accessories; I will furnish only a dish done in my way; it will be a little sucking piglet, fattened according to my method. Two hours before the dinner, I will present it to you alive and well, you will be responsible for killing and preparing it and I will not come near it until the moment when it is served on the table. You will cut it into four equal parts, you may serve me the part that you judge to be suitable. The day after this dinner, one of four things will happen: either we will both be dead, or neither of us will be dead; or I will be dead and you will not; or you will be dead and I will not. Of these four chances I will give you three and bet you 5,000 guineas that the day after the meal you will be dead and I will be well.

Cagliostro's challenge appeared in the *Public Advertiser* of 5 September, and de Morande's reply was uncharacteristically feeble. He refused a duel, he said, because he didn't want Cagliostro's death on his conscience. However, if the arsenic experiment could be performed on a cat or a dog, he'd be pleased to try it. Cagliostro had scored a sharp blow against a man already notorious for cowardice, and he pressed his advantage home on 9 September. He replied that a cat or dog would be too superior to stand in for a beast like de Morande. Moreover, "It is not your representative but it is you that I wish to dispatch."

Unfortunately for Cagliostro, de Morande was both smarter and had better resources than dour Dr. Roggerson. The French pressman had the cunning of a weasel, a bankroll from the Bourbons, and a brand of poison in his pen for which the Copt had no antidote. His initial response to Cagliostro's jeers was to trawl minutely through the publications of the Copt's growing number of European critics. As a seasoned gutter journalist, de Morande also knew that the brazen repetition of lies was as good as the truth. Some of his most telling attacks were built around the lies of Carlo Saachi.

Gradually, de Morande's evidence against Cagliostro began to mount. Soon he got wind of a recent article published in Berlin by Elisa von der Recke, claiming that Cagliostro was a necromancer and fraud. This hurt the Copt badly. Elisa had written him several devoted fan letters after he left Courland, but in the months and years that followed she'd gradually moved into the enemy camp. Masons in Mitau had disillusioned her with stories that Cagliostro had bribed his *pupille* to perform. Chancellor von Howen confessed that he'd lent Cagliostro a large sum of money that was never repaid. She heard some bad reports, too, from Russia and Poland. To clinch matters, Elisa had been courted during the early 1780s by German intellectuals who wanted to enlist her in a war of reason against Masonic superstition. Belting Cagliostro became her path to literary glory.

Tracking Cagliostro's movements through Europe, de Morande gathered letters from correspondents in Russia who claimed that the empress had expelled the charlatan for swindling, and from Warsaw saying that Cagliostro had been caught there swapping crucibles. De Morande's legal training also helped him to sniff out the trail of affadavits and counter-claims that Cagliostro had left behind in London a decade earlier. These documents appeared in the *Courier* one after the other. In particular, Cagliostro's use of the name Colonel Joseph Pellegrini of the Brandenburg army looked bad.

De Morande's knowledge of émigrés' haunts also helped him find witnesses like an Italian restaurateur who remembered Cagliostro from the Italian dives of Soho in 1772, when he'd been called Balsamo. Cagliostro's refusal to admit he'd ever made this first visit to London disclosed another chink that the pressman was quick to exploit. Gradually his articles in the *Courier* focused on collecting proofs that all three visits to London had been made by the same man, and that Balsamo, Pellegrini, and Cagliostro— three imposters—were one and the same person.

Shrewd tactician that he was, de Morande realized that a special campaign was needed to undermine Cagliostro's standing in the Masons. In particular, the Copt's declared intention of founding a lodge of Egyptian Free-masonry in London had to be stopped.

This was no easy matter, because Cagliostro's stock suddenly rose in September 1786 with rumors that he'd won a significant new Masonic ally. Philippe Jacques de Loutherbourg, an Alsatian artist, was as distinguished as Gordon had been dodgy. The famous philosopher Denis Diderot had once described de Loutherbourg as a genius, a verdict which the French Academy of Painting endorsed by electing him in 1767 as its youngest fellow ever. After he'd migrated from Paris to London four years later, de Loutherbourg gained even greater fame in the theater world. David Garrick, the actor-manager of the Drury Lane theater, offered him a handsome salary to transform the dullness of British stage effects. Over the next fourteen years, de Loutherbourg had become a maestro of spectacle, introducing brilliant innovations in set design, lighting effects, automata, and sound technology.

These talents, combined with expertise derived from his early training as an engineer, had also enabled de Loutherbourg to make artistic and scientific advances in the field of illusion. During the early 1780s, he'd perfected a small-scale moving-image show called the Eidophusikon that took the London art world by storm. Within a miniature six- by eight-foot stage, he'd developed a series of breathtaking visual transformations, using lifelike mechanical automata and naturalistic light and sound effects. Its most famous set piece had been a dramatic version of Milton's "Pandemonium." Minature mechanical models of Satan's diabolical hordes erupted from the underworld of Hades, surrounded by black boiling lava, sulfurous yellow smoke, and dancing blue lights, to an accompaniment of eerie sound effects.

Shows using ghostly special effects were, in 1787, to be given the name of "phantasmagoria," but de Loutherbourg actually pioneered the form six years earlier when a rich young aesthete, William Beckford, asked him to pour "the wildness of your fervid imagination" into creating an occult eastern spectacle at his country house. De Loutherbourg's "necromantic" light effects, "preternatural sounds," voluptuous scents, and clockwork machinery had so intoxicated young Beckford that he'd immediately begun writing his famous Oriental romance, *Vathek.* He didn't know it, but his imagination had been seized by the forerunner of the modern cinema.

Newspapers were puzzled about why a rich and talented society darling like de Loutherbourg should choose to befriend Cagliostro, particularly after the recent bad publicity associated with Gordon. The painter himself explained his new connection with Cagliostro as a simple business partnership: he and the Copt had agreed to share the proceeds of their future work—de Loutherbourg would sell his paintings and Cagliostro his medicines.

De Morande sensed that there was more to the relationship. For a start, the Alsatian had gone to the university in Strasbourg, the city where Cagliostro had lived for three years, working as a Mason and healer. The journalist probably guessed that the two men had met there — in fact, they had, at Strasbourg's Amis Réunis Lodge in 1783.

Moreover, de Loutherbourg, like so many foreign-born artists, was notoriously addicted to the mysteries of alchemy. He'd assembled one of Britain's most extensive occult libraries, and he also ran a well-equipped laboratory. It was rumored that he'd spent so much time experimenting in the laboratory that his new wife, Lucy, smashed some crucibles in exasperation. One could see why such an ardent alchemical seeker would be drawn to Cagliostro. De Loutherbourg boasted in 1786 that he'd seen the Copt transmute copper pennies into silver and cure the sufferings of his arthritic neighbor in Chiswick, Elizabeth Howard. Like Swinton, the painter obviously hoped to profit from Cagliosto's alchemical skills.

Cagliostro could also help market a more recent medical craze. A few years earlier the de Loutherbourgs and their arty friends Richard and Maria Cosway had paid lavish sums to attend a series of lectures on animal magnetism by a doctor, John Benoit de Mainauduc, who had been trained in Paris. They'd learned how a mesmerist's hands could stroke the body free of the blockages that caused physical and psychic illnesses by preventing the healthful flow of invisible and imponderable fluid. Cagliostro, of course, believed himself a medical genius superior to all others, and he hinted at a profound knowledge of this fashionable European science.

And of course de Loutherbourg was excited by the Copt's mystical Masonic agenda. Though the painter had for social reasons enrolled in London's respectable Grand Lodge, he'd secretly joined a number of smaller, more heterodox lodges and societies in England and abroad. To a theosophical groupie like de Loutherbourg, Cagliostro was the biggest star of the mystic international ever to visit the country. Also, the timing of the Copt's visit could hardly have been better. With a little help from a friend, he could present himself as the natural successor to Falk and Swedenborg; both had died in recent years, leaving a scattering of lost and leaderless disciples.

De Morande, of course, found it easy to explain why a charlatan like Count Cagliostro should want to ally himself with a luminary like de Loutherbourg. A recent tract, *Cagliostro Unmasked in Warsaw,* had ridiculed the charlatan's clumsy attempt to perform alchemical transmutations at Wola in 1781. The author (Count Moczynski) had jeered that if Cagliostro "was a little more versed... in the optical, acoustic, and mechanical arts, and in physics in general, if he had studied even a little of the tours of the magicians Comus and Philadelphus, what a success he could have made for himself." Cagliostro had obviously taken the tract's advice to heart: he'd linked up with the most highly skilled illusionist in the world.

De Morande warned the *Courier's* readers that Cagliostro would soon be using illusionist tricks to bewitch his disciples in London. Mind-bending techniques like these had been used by a fraudulent Mason in Leipzig, Johann

Eight Masonic studies by de Loutherbourg intended for Cagliostro's lodge of Egyptian Masonry

Georg Schroepfer, who'd later blown his brains out when his deceptions were discovered. Hidden magic lanterns with magnifying lenses would play on moving transparencies to produce wraithlike images of the dead surrounded by tinted clouds. The virtuoso de Loutherbourg would find no difficulty producing such haunting spectral effects.

For once, de Morande was wrong. Cagliostro actually intended to use more traditional aspects of his partner's genius to help propel the Egyptian Masonic mission. In the autumn of 1786 he persuaded de Loutherbourg to produce a series of paintings designed eventually to hang in the initiation chambers of a proposed new lodge of adoption in London.

Following detailed instructions from the Copt, de Loutherbourg painted a set of eight Masonic watercolors. They evoked in symbolic form the ordeals that an initiate of Isis had to overcome before reaching the ultimate level of Egyptian Masonry. As she entered the lodge, the candidate would encounter the first painting, which represented both the goal and the struggle that lay ahead. It showed a sensuous young woman dressed in a white robe with a blue sash—an idealized version of Seraphina—gazing longingly at a

holy city high in the clouds, while an angry green serpent reared up in her path.

In the second painting, the candidate has successfully advanced to the next degree, symbolized by a ruined classical temple with treasure strewn on the floor. Her right hand holds a sword which has just severed the satanic serpent's head, while her left hand is clasped to her breast.

The third painting shows her contemplating a new level of challenge. She stands with downcast eyes in front of the Grand Maîtresse, who wears an owl-adorned helmet like that of Pallas Athena and points to a horde of androgynous witches waiting in the path. These are both sexual temptresses and monsters of depravity.

By the fourth picture the initiate has reached the foot of the temple — an Egyptian pyramid surmounted by a vivid rainbow. It is a moment of near-despair. She sits on a rock, exhausted, broken weapons strewn at her feet, while witches with pendulous breasts and snakes twisting from their genitals caper in her path. They brandish blazing torches and handfuls of snakes. In the foreground, however, stands the Great Copt himself, sword in one hand,

miter in the other, urging her toward the holy arcanum at the front of the temple.

The next painting confronts her with humanity's greatest test—mortality itself. In front of a cave below the temple stands a giant muscular figure of Father Time; he is winged and bearded with an hourglass perched on his head and the scythe of death held in his right hand. His intentions are obvious.

Just when the prospect of spiritual ascension seems insurmountable, the Great Copt comes to the rescue. He is represented as a trimmed-down version of Cagliostro in a red sash. The Copt grabs Father Time by the wings and prepares to lop his feathers with a two-edged sword. In the foreground lies the upended hourglass; behind him a phoenix rises triumphantly from a blazing pyre. Time, the enemy of man, has been conquered.

But the Copt must fight and kill another opponent as well—a Hermes figure with winged heels. Presumably this is not the original Hermes Trismegistus from whom Egyptian Masonry derived, but the false Mercury whom orthodox Masons follow. The two grapple, and Cagliostro strikes Mercury to the heart with his sword.

Finally, de Loutherbourg offers the brave pilgrim the ultimate reward of her endurance. She has arrived at the steps of the Egyptian temple, and her consecration is moments away. Beckoning, the Copt gestures with his sword toward the remaining few steps she must ascend. A soft blue veil pierced by beams of radiant light flutters over the threshold. On the stone walls, fragments of words suggest the deepest occult mysteries—ARCANUM MAGNUM; GEMATRIA; SECRETS. To the pilgrim's right, the false Mercury lies prone—dead or sleeping—after his defeat. He slumps over a headstone carrying the words, PIERRE BRUTE, signifying the gross material concerns of Cagliostro's Masonic rivals.

It's unlikely that de Morande ever saw these paintings, but he worked out Cagliostro's intentions anyway. A canny political analyst, he didn't need to be told of the Copt's ambition to lead British mystical Masonry. And as a respectable Modern Rite Mason himself, he knew how to persuade his brothers to reject this dangerous foreign pretender. He proceeded to attack Cagliostro's Egyptian mission on multiple fronts. First, he ridiculed the Copt with a burlesque description of the initiation ceremony at the King's Head a decade earlier. The idea of a tubby Italian crashing from the ceiling onto the hard tavern floorboards, and then blubbering with terror as an empty pistol was put to his temple, raised guffaws on both sides of the Channel.

De Morande realized, too, that it was crucial to exert particular pressure on the Swedenborgians of London who were most likely to be receptive to Cagliostro's mission, so the *Courier* invented a story that was to become legendary. The Copt had supposedly given permission to a group of local Swedenborgians to use his secret rituals to evoke angelic spirits. But "in the place of Seraphim in azure and silver robes that they had hoped for, there appeared a terrible mob of wild orang-utans whose grimaces, insults and unworthy promiscuity...the chaste idealists had to endure all evening." The mirth of de Morande's readers, as they contemplated the high-minded

Swedenborgian séance thrown into chaos by a pack of obscene apes, impelled Dr. Chastanier, the local leader of the society, to warn his flock to steer clear of Egyptian Freemasonry.

Cagliostro's mission received another devastating setback at the beginning of November. A few days earlier, the Copt and his closest émigré disciples had been introduced to a meeting of the Antiquity Lodge number 2, held in King Street, Bloomsbury, at the house of a *perruquier*, Mr. Barker. Cagliostro intended a takeover but, according to de Morande, the overture turned quickly into farce. An artisan called "Brother Mash," who was to conduct a welcome ritual, performed a tavern-style burlesque that caricatured Cagliostro as a fairground quack.

Despite this unpromising rehearsal, Cagliostro launched his formal bid for leadership the following day. The *Morning Herald* of 2 November carried an advertisement appealing to "all true Masons" to combine with Swedenborgians to regenerate British Freemasonry. Using the mystical numerical codes beloved by initiates, it urged that "the building of the new Temple of New Jerusalem, 3, 8, 20, 17, 8 [church]" should begin. A grand meeting was summoned for 3 November at O'Reilly's tavern on Great Queen Street, where a foundation stone of the new lodge would be laid. But the same day the *Courier* ran de Morande's scathing description of the fiasco in Bloomsbury and Cagliostro's appeal fell on deaf ears.

A week or so later, the Copt experienced another humiliation when a witty, graphic caricature of Brother Mash's burlesque hit the London streets. It was engraved by James Gillray, a satirist of towering genius, who was temporarily in the pay of the British government. Entitled "Extract of the Arabian Count's Memoirs," Gillray's engraving showed the portly Cagliostro in his Masonic regalia trying to tout his drops to a skeptical group of British brothers who jeer at the charlatan and his entourage of foppish-looking foreign disciples. A pungent verse caption explained the story.

Gillray's satire on Count Cagliostro's attempt to found an Egyptian Masonic lodge in London

Born, God knows where, supported God knows how,
From whom descended—difficult to know;
Lord Crop adopts him as a bosom friend,
And madly dares his character defend.
This self-dubb'd Count some few years since became
A Brother Mason in a borrowed name;
For names like Semple numerous he bears,
And Proteus-like in fifty forms appears.
"Behold in me (he says) Dame Nature's child
Of Soul benevolent and Manners mild,
In me the guiltless Acharat behold,
Who knows the mystery of making gold;
A feeling heart I boast, a conscience pure,
I boast a Balsam every ill to cure,
My Pills and Powders all disease remove,
Renew your vigour and your health improve."
This cunning part the arch-imposter acts
And thus the weak and credulous attracts.
But now his history is render'd clear
The arrant hypocrite and knave appear;
First as Balsamo he to paint essay'd
But only daubing he renounc'd the trade;
Then as a Mountebank abroad he stroll'd;
And many a name on Death's black list enroll'd.
Three times he visited the British shore,
And every time a different name he bore;
The brave Alsatians he with ease cajol'd
By boasting of Egyptian forms of old.
The self-same trick he practis'd at Bourdeaux,
At Strasbourg, Lyons and at Paris too.
But fate for Brother Mash reserv'd the task

To strip the vile imposter of his mask.
May all true Masons his plain tale attend!
And Satire's laugh to fraud shall put an end.

As Casanova pointed out in his own attack on Cagliostro published in Warsaw the same year, charlatan-magicians could cope with anything but ridicule. Cagliostro did his best. With the help of Thilorier, who was in London again for legal consultations, the Copt tried to retaliate against de Morande's jeers by publishing a *Letter to the English People*. Far from denying that he'd visited the country in 1776–1777, he explained that he'd been victimized at the time by a gang of swindlers similar to Swinton and de Morande. Both of the latter, he went on to argue, were grubby extortionists who'd turned on him when he resisted their demands.

In support of his case, Cagliostro cited some of de Morande's notorious blackmailing episodes of the past. This French scoundrel had been bribed by the French ministry to perjure himself. He was also a notorious coward who'd slithered out of duels with his victims and successively betrayed every one of his friends. The *Letter to the English People* praised the liberalism and generosity of Britons who'd protected Cagliostro from French tyranny, and it concluded with the Copt's customary reminder of the terrible fate that had befallen all those who'd crossed him in the past.

Ink, however, was not the only type of poison that de Morande could apply. Behind him stood the power of law wielded by the British and French governments. In the winter of 1787 he crowed that a series of prosecutions had been launched against Cagliostro's friend Lord George Gordon. The incendiary had been charged on 27 January 1787 by the British and the French governments with separate counts of libeling the British judiciary and the French queen. Soon, he disappeared into hiding.

As a former lawyer, de Morande also knew a subtler way of using the British law for counterrevolutionary ends. Though Cagliostro could not himself be charged with libel, he could be made a tenant of the debtor's section of Newgate prison. De Morande incited half a dozen creditors to come forward with claims new and old. Bailiffs were hidden in Swinton's house several times, intending to seize the count; luckily Cagliostro received enough warning to post bail and stave off these legalized attempts to kidnap him. Undeterred, de Morande paid the costs of bringing the predatory druggist Carlo Saachi over from Strasbourg to sue Cagliostro for 150 pounds of unpaid wages.

Next, he announced that he was arranging to bring over a Spanish gentleman, M. Silvestre of Cadiz, who'd been pursuing Cagliostro for years over the cost of an ornate silver cane. The same accusation had embarrassed Cagliostro long ago in Saint Petersburg. Seven years later, it proved to be the straw that broke the camel's back. Realizing that his freedom and all their costly furnishings in Knightsbridge were at risk, Cagliostro used the only magic trick left in his stock. At the beginning of April 1787 he suddenly vanished to Switzerland, prophesying the imminent death of his remorseless attacker.

Théveneau de Morande responded with a declamation of his own. The *Courier de l'Europe* of 6 April 1787 carried the simple headline: HE IS DEAD — announcing not only that Cagliostro's European reputation was extinct but also that the Copt's Masonic disciples in Britain had abandoned their prophet.

· 6 ·

Rejuvenator

Rejuvenate: To restore to youth, to make young or fresh again.

On 17 June 1787 a heavily laden English coach crossed the borders of
Germany and France to enter the Swiss town of Bâle (Basel). It was a lovely
prospect for the passengers, Seraphina Cagliostro and Philippe and Lucy de
Loutherbourg, as they went up the steep cobbled hill toward the twin
mansions of the Sarasin brothers—Swiss silk merchants, bankers, and tex-
tile producers. Below, to the left of the carriage, lay the broad sweep of the
Rhine, its olive-green waters taking fat barges filled with ribbons, silks, and
calicoes to the trading ports of Europe, the Levant and the Indies.

Religious strife in the seventeenth century had driven families of skilled
Huguenot silk workers like the Sarasins to the haven of this quiet old river
port and city-state. A century later their entrepreneurial energy and judicious intermarrying had transformed Bâle into a thriving merchants' and
bankers' oligarchy of some 15,000 inhabitants. The town's broad streets,
town houses, and steepled churches reflected its reputation for prosperity,
order, and restrained Protestant piety.

The coach was heading for the house of Jacques Sarasin and his wife, Gertrude, whose life Cagliostro had saved in Strasbourg six years earlier. The route wound alongside the river and up a steep road called Martingasse past the neoclassical facade of Lucas Sarasin's "Blue House" to reach their destination, its exact twin, the "White House"—a bland name for a building that had cost more than 200,000 guilders and occupied a full block. It looked like a gleaming ivory palace set off with a crown of fine-meshed orange roof tiles. Over the studded front door two plaster satyrs offered alternative welcomes; one grinned lasciviously, the other bared its teeth in an angry grimace.

Inside the coach Seraphina Cagliostro, about to see her oldest and dearest friends, the Sarasins, was more anxious than usual—she was fretting about the presence of their houseguest, her husband, Count Cagliostro. As the coach turned into the cobbled rear courtyard, she glimpsed: Jacques Sarasin, his heavy Swiss frame and solemn demeanor softened by a cascade of chins; and Gertrude, round and ruddy with pregnancy. But were they smiling or frowning? More crucially, what kind of welcome would she get from the mercurial figure at their side?

Seraphina had good reason to be nervous. During the three months since she'd last seen Cagliostro in London, she'd betrayed him by blurting out intimate and damaging secrets to his deadly enemy, Théveneau de Morande. She had not really intended to, but she'd been utterly fed up with her husband. He doted on her, she couldn't deny that; when flush with funds he showered her with gifts, and he grew inconsolable when they were apart. But he was also prone to moodiness and bouts of explosive rage, particularly when things weren't going well, and lately they hadn't been.

Seraphina had rebelled suddenly against their whole way of life. She hated their sudden oscillations of fortune, particularly those times when she was forced to dismiss her servants and pawn her jewels. She was fed up with the forced moves that tore her from lovers and friends. Then there were the humiliations inflicted by the police, bailiffs, and turnkeys. Above all, she disliked living in Britain; the language baffled her, the religion left her cold, and the weather made her miserable. She longed for the warmth of Italy, the style of France, and the comforts of Catholicism.

A year ago in London her husband had told one of the Sarasins' friends that it was only the love of his wife that stopped him from flying into the African jungle to live with the wild beasts. Yet the following year he'd fled to Switzerland, leaving his wife to deal with beasts: creditors trying to seize her furniture; servants suing for back wages; Masons from France bombarding her with questions. Was Count Cagliostro, they asked, really the crooked Sicilian Giuseppe Balsamo? Did the European network of Egyptian lodges exist outside his imagination? She'd borne the brunt, too, of the anger and hurt of Cagliostro's idealistic young secretary, who'd caught her husband fabricating an experiment in London just before the Copt ran off to Switzerland. The young man told her that he was furious at having wasted almost a year of his life studying with a charlatan, and he was so depressed that she offered him sexual consolation. And why not? She was alone, and he was ardent.

To her surprise, the journalist Théveneau de Morande proved equally charming, despite his terrible reputation. When he heard of Cagliostro's departure, he rushed to visit her with an offer of help. She didn't intend to talk to him quite so freely, that's true; but he was difficult to resist. He sympathized with her loneliness and her discontent at the vagabond life Cagliostro imposed on her. He seemed to understand her hankering for her religion and dislike of her husband's impieties. He even knew about the humiliating way Cagliostro had shut her in a French convent long ago. And it was not as if she told the man anything he didn't already know. She merely confirmed some of his speculations about her husband's early life.

Besides, she might easily have betrayed Cagliostro more seriously. French officials in London offered her a handsome pension from King Louis if she would swear to the falsity of her husband's case against de Launay and

Chesnon. She was tempted, of course, but also nervous—it might be a trap. The Bourbons were known to be unreliable. In the end, she was persuaded out of the idea by her hosts in London, the de Loutherbourgs, who also promised to take her to Switzerland, where they intended to continue working with Cagliostro. They would help her to explain to Cagliostro how she'd been cajoled by the pressman.

Seraphina need not have worried. Cagliostro greeted her with unabashed joy. He clasped her in a powerful hug and whispered his usual endearments—*cara figlia*; how he'd missed his little countess. Then he turned and embraced the de Loutherbourgs. He would never forget their kindness in delivering his darling wife. Be assured, he told them, Count Cagliostro would repay them abundantly.

Actually, Seraphina hadn't seen her husband in such good spirits for years. For the next few weeks he fairly crackled with warmth and energy. She didn't even have to explain away her betrayal: Cagliostro was convinced that she'd been tricked by disciples to whom he'd given his love and trust. They'd been bribed by the Bourbons, he said, because the French king and queen feared him above any man, and rightly so.

His darling Seraphina must not worry her pretty head, Cagliostro whispered in his most caressing manner; he understood how this pack of brigands had hounded her. He knew, too, how sad she always felt when deprived of her furniture, jewels, and servants. All would be well now. He merely asked that she make a legal declaration explaining how spies and smut-mongers had forced her to tell lies about her husband. Such a declaration could be made in front of a local magistrate and would greatly please their true disciples in Switzerland and all over the world.

Seraphina was relieved to accede to this request, and, as so often in the past, Cagliostro soon swept away her remaining doubts on a surging tide of rhetoric. It had been a mistake, he said, to go to England, a parochial and money-obsessed country whose Freemasons lacked spirituality. The Cagliostros always knew that London swarmed with predators and spies. Every time they visited, things went wrong. He, Cagliostro, should have known better than to go back there.

In retrospect, he conceded, they ought to have moved to Switzerland long ago. He'd been tempted on his first visit to Bâle, soon after curing Gertrude Sarasin in 1781–1782. There had been good reason to stay. Jacques, eager to show his gratitude, had insisted on building Cagliostro a beautiful Chinese-style pavilion at the village of Riehen, just outside Bâle—now that was the act of a true disciple! The pavilion was designed to Cagliostro's minute specifications, so that he could undertake the highest and most demanding goal of Egyptian Masonry—spiritual and physical rejuvenation. Jacques had spared no expense. True, some repairs were needed now, but Jacques had promised to attend to these at once. The pavilion had an exquisite miniature tower topped by a golden weather vane; bells, with little gilt clappers, each suspended from a corner of the tiled roof; a sumptuous meditation chamber; a billiards room; and a French-style *jardin* for soaking up nature. Back when it was built, Cagliostro admitted, he'd been too wrapped up with his clinic in Strasbourg to appreciate such a wonderful gift.

Now, he promised Seraphina, everything would change. In Switzerland, he would at last have the leisure to perfect his system of rejuvenation. This would be his new mission. He'd decided to give up trying to run free healing clinics for the poor—it only aroused the jealousy of doctors. Nor would he waste time trying to reform the hopeless corruption of orthodox Freemasonry. In Switzerland, he assured her, all this was unnecessary. Here in Bâle he was treated like a returned divinity by a brilliant circle of admirers, all friends or relatives of the Sarasins: the Hasenbachs and the Bishoffs, the French pastor Touchon and the German pastor Schmidt, Professors Haas and Breitlinger, Jacques Sarasin's brother-in-law de Gingin, the wealthy intellectual merchants Jacob Burckhardt and Isaac Iselin, the famous blind poet Pfeffel, and many others.

So impressive was the spiritual level of this circle, Cagliostro told her, that he'd already decided to establish it as the mother lodge of Egyptian

Freemasonry for the Helvetian states. This lodge, he believed, would be the kernel for the spread of his Rite throughout Switzerland. Members gathered regularly in the White House in a room free of the extravagant clutter of the Parisian lodges. Such was the simple natural piety of these Swiss: they needed only to contemplate a bust of Cagliostro in order to reach a state of spiritual transcendence.

Now, Cagliostro assured his wife, he would concentrate on caring for their dear friends and hosts Jacques and Gertrude. When the Sarasins could find the time, he would administer his strenuous forty-day course of moral teaching and meditation to cleanse them of original sin until they reached a state of complete purity. After that, he would prepare them to undertake the ultimate ordeal of physical rejuvenation; this entailed a further forty days of fasting and isolation in the Chinese pavilion. Jacques and Gertrude would have all impurities driven from their bodies by his special distilled waters and laxative herbs, and their putrid humors would be purged by bleeding. After a further dosing with grains of the universal panacea based on his secret recipe, they would convulsively shed the skin that encased their old corrupt forms. Finally, they would be reborn as immortals.

He told Seraphina that he'd already taken the Sarasins through the first steps of this future transformation by dictating to Gertrude some of his secret prescriptions, including those needed for her imminent ordeal of childbirth. Despite Gertrude's buxom appearance, everyone in the White House had been worried that she could relapse into the nervous condition from which he'd rescued her in 1781–1782. Now, thanks to his prescriptions, she and Jacques would be able to make an "elixir for preparing for childbirth," "powders to give appetite," "oil against hysterics," "elixirs for when the womb comes out of the body," and "herb tea" and "water of life" purgatives. Naturally, Gertrude was also thrilled to have his oils and powders for rejuvenation of the skin, teeth, eyes, hair, hands, feet, lips, and nose. He'd even disclosed to the couple his famous "paste of paradise" and

"Wine of Egypt" so that they could recover their normal sexual appetite soon after the exhaustion of childbirth.

In return, Cagliostro told Seraphina excitedly, Jacques had found them a wonderful spot to live. He'd settled on Bienne, a small town close enough for regular visits to Bâle, whose restrictive immigration rules prevented foreigners from settling there. Crucially, Bienne had a good climate. Built on the edge of a crystal lake looking toward the Alps, it was sheltered from glaciers and from the worst winter winds.

Moreover, friends of the Sarasins, a large family called the Wildermetts, ran the town. Banneret (Seigneur) Sigismund Wildermett had found a perfect house for them. The banneret guaranteed, in fact, that they would experience "more safety and tranquillity than in any other country of Europe." Freemasonry was warmly welcomed in the town, and there were several lodges already operating there. Sigismund's brother, Alexander, the mayor of Bienne, had also written to express delight that people of the Cagliostros' eminence should grace Bienne. He believed that the Copt's reputation as a miraculous healer would fill the town's hotels with wealthy visitors from all over Europe.

Cagliostro promised that Seraphina would love their new home. It was called Rockhalt. A steep red roof and whitewashed tower graced by a simple cross gave it the look of a country château, yet the house was situated at the edge of town on the Pasquart promenade. The main building, where they would live, had eight beds, while a smaller annex, perfect for Philippe and Lucy de Loutherbourg, had a further four. The rent would be paid by Jacques Sarasin, who offered to give them anything they needed. Nor was Jacques alone in his generosity: Rockhalt had been cleaned and freshly painted from top to bottom at Banneret Wildermett's expense. It was already well furnished, also included linen, porcelain, and cutlery, and Seraphina would have plenty of fun redecorating.

The grounds of Rockhalt were as charming as the house itself. Cagliostro waxed lyrical describing them: a wooden stable with a cow for milk, a swath of terraced land running down to a fast-flowing river, a flourishing orchard planted with grapevines, and excellent new fencing throughout. The good banneret had thought of all their needs; he'd hired a gardener, a cook, a hairdresser to look after Seraphina, and a capable local chemist to make up Cagliostro's recipes.

Seraphina would be surprised, too, Cagliostro assured her, at the liveliness of local society in Bienne. The banneret described the social life of the town as "simple and gay." And, of course, they'd have the glamorous de Loutherbourgs as boon companions. Cagliostro had also prepared a special surprise for her. Marquise Branconi, former mistress to the prince of Brunswick—the same woman who'd snubbed Seraphina seven years ago in Strasbourg—well, now she was begging to be reintroduced. Cagliostro had magnanimously agreed, provided the rapprochement took place at someone else's house. He was certain that the dazzling marquise would make Seraphina an exciting playmate.

Sure enough, when they finally arrived in Bienne at the end of June, the banneret gave a lavish welcome party attended by local dignitaries, including Marquise Branconi accompanied by two bickering lovers. The elegant marquise was as effusive as she'd once been snooty, and Cagliostro pleased the locals by saying that he'd at last found the peace he'd been seeking for so long—hadn't he told Gertrude Sarasin that he longed to "live in philosophical repose, irreconcilable with the tumult of the world"? The de Loutherbourgs also made a good impression, though Seraphina, "radiant with diamonds," was flattered that everyone thought she was the most beautiful woman present, notwithstanding the presence of the English-woman. Cagliostro, needless to say, was lionized by society matrons clamoring for his famous beauty and nerve treatments.

After only a few weeks, however, Cagliostro's mood began to darken. Despite the isolation of Bienne, news trickled in through local friends that all over Europe former acquaintances were baying after the count like savage mastiffs. The diamond necklace affair, combined with de Morande's attacks, had evidently created a lively demand for attacks on the famous Mason — all the more so because de Morande claimed to have nosed out documents from the archives of the police in Paris containing samples of handwriting supposedly proving Balsamo and Cagliostro to be the same man. The hunting season on the Copt was now open.

The older Seraphina

One of these hostile memoirs, a small anonymous tract called *Soliloque d'un penseur* published in Warsaw early in 1786, was obviously from the pen of their adventurer acquaintance Casanova. After leaving Venice in 1781, when the Inquisition Tribunal finally lost patience and sacked him, he'd taken a position as librarian in a Bohemian castle. Here he'd decided to recapture some patronage by fashioning his encounters with Giuseppe Balsamo into a "philosophical" attack on the European vogue for magical healers.

Soliloque had also benefited from Casanova's chance encounter with another former friend of the Cagliostros, Elisa von der Recke. In the winter of 1785–1786, Casanova shared a dinner table with her at the fashionable Austrian spa town of Teplitz. Elisa was taking the famous waters, Casanova escaping from the tedium of the castle. The old roué and the earnest pietist took to each other: he adored beautiful woman and she admired older men. Casanova might be patched and powdered, but he was still a formidable charmer. Elisa proved to be literally his last conquest, though his old volcanic lusts were now directed toward writing flowery letters on the soul. And, of course, Elisa and he had shared a common project.

Elisa's *Report on the Famous C. at Mitau in* 1779, published in Berlin in 1786, was making a far greater impact than Casanova's tract, not least because it came from the pen of a former disciple rather than a jealous rival. It took the form of her original excited journal of 1779 with later annotations supposedly explaining how Cagliostro had managed to bewitch her. This recipe proved a resounding success. Everyone wanted to read it. Elisa personally discussed the work with many of Germany's most famous intellectuals—the philosopher Immanuel Kant, the musician Carl Philipp Emanuel Bach, the poet Friedrich Klopstock, the writer Johann Wolfgang von Goethe, the moralist Johann Georg Hamann. No wonder news of the *Report* reached even the remote Swiss hamlet of Bienne. The Cagliostros had been in the town only a few weeks when Sigismund Wildermett revealed as tactfully as he could that Elisa's memoir was said to be turning all the German states against Cagliostro.

Cagliostro also found himself having to cope with a flood of attacks from farther east. Count Moczynski's exposé in 1786 of Cagliostro's alchemical bungle at Warsaw was damaging enough, but much more serious was news that the empress of Russia was mounting a personal literary campaign against him. Cagliostro's visit to Saint Petersburg had begun a change in her attitude toward Freemasonry that intensified after 1784 when the Bavarian Illuminati were outlawed for operating a revolutionary and atheistic underground. Her vague suspicions of Cagliostro's Russian mission hardened into steely certainty.

In 1785 Catherine wrote to one of her most trusted advisers, outlining a plan to launch a European-wide mission against Freemasonry using Cagliostro as her target:

> The spirit has come to me...to address a manifesto to my people in order to put them on guard against the seductions of a foreign invention, the Masonic lodges.... They tend to destroy Christian orthodoxy and all government, and from the place at which they were born

comes disorder under the form of a pretended equality which does not exist in nature. At the same time they foment all the crimes against human and divine laws of all world civilizations, renewing pagan ceremonies, evocations of the spirits, research for gold or for the universal panacea.

The first fruits of this mission were two new satirical plays, *The Tricker* and *The Tricked*, written one after the other in 1785. Catherine told her informal foreign agent, Dr. Johann Zimmerman, a cool and ingratiating Swiss intellectual living in Germany, that they represented "Cagliostro in nature...and his dupes." The first play, *The Tricker*, showed him as Kalifalk-gerston, a pompous swindler who cheats the innocent by pretending to make their pearls and diamonds grow. The second play, *The Tricked*, presented a similar story from the perspective of a dupe, Radotov, who, after being initiated as a Mason, begins to gibber nonsensical phrases, believes he can speak with spirits, and tries to transform metals into gold. By the time Radotov's wife and niece rescue him, he is a cretin.

Both plays were staged at Catherine's theater in the Hermitage in January and February 1786, then immediately afterward at the national theaters of Saint Petersburg and Moscow. Catherine also took the fight to what she believed were the sources of infection—Germany and France. She allocated six hundred rubles for staging them in Hamburg and Paris. Zimmerman also sent German translations to a professor at Göttingen who published them in various local journals.

Catherine wrote her most ambitious attack on Egyptian Masonry while Cagliostro was still in London. *The Shaman of Siberia* had more weight than her earlier satires because she'd just read Elisa von der Recke's *Report* and was so impressed that she ordered it to be translated into Russian and circulated throughout her dominions. She also modeled a key character in the play on Elisa. A young girl, Prelesta, is afflicted with an unknown nervous disease, so her gullible father summons a famous Masonic shaman, Amban-Lai, to heal her. (Catherine's audience knew that Prelesta's real problem stemmed from frustrated love due to a quarrel with her fiancé.)

Amban-Lai the shaman is both more sinister and more sincere than Catherine's earlier versions of Cagliostro. Her reading of Elisa led her to see the Copt as a modern version of the fanatical religious mystics of the Russian church. Her Cagliostro figure is also much more countercultural, a deluded visionary who wants to reform all "science and manners." He is "an apologist of silence, of the suspension of the senses, and a detractor of philosophy and reason," who "has an aversion to educated people, to the spirit of discovery and research, and wishes to plunge everything back to barbarism, if the government encourages or allows it."

In Bienne, Cagliostro also found himself having to cope with a new flare-up of interest in the diamond necklace affair. In August 1787 newspapers all over Europe excitedly revealed that Jeanne La Motte had made a daring escape from the Salpêtrière prison, disguised as a man, to become an exile in London. She was now apparently hatching a new batch of libels accusing Cagliostro, Rohan, and the French queen of complicity in the swindle. The newspapers in Britain and France were filled with her fresh accusations.

In the same month another bombshell was lobbed into the little Swiss town from France. Newspapers reported that the law case in Paris was lost, in embarrassing circumstances. The king had savaged the count's lawyers for wasting the court's time with a trivial and vexatious plea intended to swindle the defendants. This was the most humiliating blow of all: de Loutherbourg wrote a concerned letter to the banneret, asking him to make a special effort to console Cagliostro, who was devastated. The company at Rockhalt, the painter said, now feared that the bastards in Paris would follow up their victory by pursuing the Copt with claims for damages.

Ironically, it was actually de Loutherbourg himself who proved to be the bastard. The painter and his wife were notoriously moneygrubbing; as soon

as they arrived in Switzerland, they began bleat-
ing about a loan advanced to Cagliostro in London. As the weeks drifted by, servants told
Seraphina that de Loutherbourg was fum-
ing about the debt and about Cagliostro's
failure to begin a promised program of rejuvenation. The Copt should spend less time
at soirées with his featherheaded wife and
more in teaching his disciples the mysteries
of alchemy, healing, and Egyptian Masonry.

Some of the Copt's friends suggested that
the real problem was sexual jealousy. Cloistered
in the miniature château of Rockhalt with little to do, the two couples and their servants
were becoming caught in crosscurrents of erotic
tension. Despite the fact that de Loutherbourg

*Philip James (Philippe Jacques)
de Loutherbourg*

looked like a fat gray turtle with a short neck and a snapping mouth, he
still fancied himself something of a libertine. He'd developed a leering
interest in Seraphina when she was staying with the de Loutherbourgs in
London, but she preferred Cagliostro's vigorous young secretary. Never
short on vanity, de Loutherbourg showed his pique by treating the countess in a patronizing manner. The rejection also inflamed his obsession with
undertaking Cagliostro's famous rejuvenation treatment, not least because
he needed to keep up with his beautiful wife, Lucy.

As for that English vixen: Seraphina confided to her friends that the
woman tried to vamp every man she met. Servants whispered to the countess that Madame de Loutherbourg was trying to seduce Cagliostro. Under
pretence of suffering from nerves, Lucy monopolized hours of the Copt's
attention. Annoyed as much by the woman's gall as by jealousy over her
husband, Seraphina confronted de Loutherbourg with his wife's behavior.
Instead of being grateful, the painter rebuffed her rudely.

In addition to these sources of tension, Seraphina and Cagliostro grew irritated at the English couple's social airs. It was obvious that the de Loutherbourgs were trying to establish a rival salon in the town. Seeing how much the banneret admired the Cagliostros, the English couple had ingratiated themselves with his brother, Alexander Wildermett. The mayor's well-known susceptibility to "the petticoat" made him a sitting duck for the vixen Lucy.

This simmering rivalry finally boiled over when Prince Edward visited the town from England. Cagliostro held the floor at a public reception, toasting the prince's mother and father as "good people," and the de Loutherbourgs encouraged Edward to view this behavior as vulgar. As a result, the prince later called on the de Loutherbourgs at Rockhalt but made no effort to see the Cagliostros. The Copt exploded: nothing infuriated him more than a social slight, and Seraphina encouraged him. Coarse remarks about a fat painter and his tart resounded through the corridors.

A week later, the Loutherbourgs left Rockhalt to move in with the mayor. The painter simultaneously issued a legal writ through the Bienne council, suing the Copt for a debt of 170 louis. Cagliostro at once filed a counter-claim. The placid burghers of Bienne squirmed at these sensational events. Banneret Wildermett wrote to Sarasin of his distress that men of such lofty and kindred principles should quarrel—this was a product, he believed, of the hothouse proximity of the two households.

De Loutherbourg tried to win over the Sarasins with letters abusing Cagliostro and whining about "his poor sick wife attacked by bile." None of it cut any ice with Jacques and Gertrude. Seraphina gathered that they were actually pleased at the rift; they'd been annoyed by the way the English couple tried to push to the head of the line for rejuvenation. Seraphina essayed a letter to them of her own, a difficult task because she'd learned to write only two years before, in the Bastille. With disarming simplicity, she begged her friends to forgive her crude writing style and expressed her sadness at the churlish behavior of the de Loutherbourgs. She embraced Jacques and Gertrude with all her heart.

Jacques was privately furious at the painter's betrayal of "our dear master," suggesting that behind a plausible manner de Loutherbourg had always masked his intention to ruin Cagliostro. Now the man's true villainy was in the open at last. In fact, he was behaving like a madman, having tried to challenge Cagliostro to a duel. The Copt had laughed scornfully, saying he fought only with arsenic. However, servants reported that the painter had sent for powder and ball and was threatening to shoot Cagliostro like a dog in the street. The Copt demanded protection by the police and the expulsion of the de Loutherbourgs from the town.

Cagliostro ought to have left it there, while sympathy was still on his side, but his Sicilian blood was up. Against Seraphina's advice, he began making wild accusations—that Lucy de Loutherbourg had called him a poisoner; that the mayor was involved in a plot to kill him; and, worst of all, that all Swiss people were naturally mean and treacherous. Realizing that the affair was escalating into a scandal that could damage the town, the banneret begged Jacques Sarasin to help. The Swiss banker dashed to Bienne on 10 January 1788 to rescue "our dear master from the claws of his enemies." In four days of tough bargaining, Sarasin forced an agreement that he believed gave Cagliostro the moral advantage, while conceding a minimum settlement to satisfy the painter's legal case. Given that Jacques also agreed to pay the latter sum himself, Sarasin believed he'd served his master well. Sarasin whisked the Cagliostros away from the overcharged atmosphere of Bienne, and for a week or two it was like old times. Seraphina was pampered by the children and servants, and the Copt renewed his course of higher instruction for local Masons. But then a decision to hold a séance using the Sarasins' youngest son, Felix, as a *pupille* turned sour. After the Great Copt initiated the boy as seer and summoned the good spirits as usual, Felix suddenly introduced an unwelcome spectral visitor to the séance:

"I find myself in a dark room. I see a golden sword suspended above my head," said the little boy. "Oh, I perceive Loutherbourg arrive. He opens his breast, and shows a wound in his heart; he holds out a poniard to me."

Gertrude and Seraphina, badly shaken, confronted Cagliostro afterward: they were worried about the impact of this gory apparition on little Felix. It must have been a diabolical vision, surely? It was too macabre to come from the great God. Cagliostro replied that the archangels were merely using the child to confirm the fate that awaited the treacherous painter; he'd either go mad or die, as Cagliostro had often predicted. Seraphina, who knew Cagliostro's black moods better than anyone else, could see that his hatred of the painter was festering. Then suddenly, by some strange twist of passion, Cagliostro began berating his hosts, muttering that Jacques had let the knave off too lightly; it showed that the banker could not love Cagliostro with his whole heart.

By the time the Cagliostros returned home to Rockhalt a few weeks later, Seraphina was sick of Switzerland and worried about the consequences of the silly affair. She longed to leave for warmer climes: she'd grown to hate the icy mountain winds. As the river froze and the snow thickened on the terraces at Rockhalt, even tough Cagliostro had to treat himself with bleedings and purgings to offset the effects of chills. He seemed feverish and distracted in a way she'd never seen before. Naturally he declared himself on the point of death.

Meanwhile, sly de Loutherbourg had developed a new goad. Helped by a printer, the painter circulated around the town two small satirical engravings that inflamed Cagliostro worse than his fever. Knowing that the Copt had administered emetics to patients suffering from hysteria, the artist depicted a roomful of visitors at Rockhalt, vomiting explosively after drinking Caglistro's famous "Eau de Minerale."

The second caricature was crueler still: it showed the tubby Copt, dressed like a fairground mountebank, summoning his famous seven spirits. Resembling monkish specters rather than angels, each carried an emblem of Cagli-

Composition latin âla, Cagliostro
Maus melas Récolas Sou Coumedt gougomerna
cus et pilloulla Dora, tousana famasavomontiva

Depence pour
herbage, 10

profit des herbage
à piloullé, 1000000

Boîte de famara
arcana d'algimica

Lettre des greanie
universel
900

Le Dejeune avec les Eau Minerale du Comte Cagliostro

De Loutherbourg's satire in Bienne against Cagliostro:
"The effects of Cagliostro's mineral water"

ostro's recent exploits: swindled jewels, loot from Egyptian Masonry, credits from Sarasin's bank, bogus law claims, and a vanishing reputation. In the background Countess Seraphina, Marquise Branconi, and Madame Wildermett gossiped together, fool's bells woven into their fashionable curls. They failed to notice the grim reaper seated in the bushes at their side.

On 23 May 1788, after de Loutherbourg had suddenly repudiated Sarasin's settlement and gone back to the law, the grand council of Bienne ruled in favor of the painter. Humiliated, Cagliostro threw "terrible scenes," accusing his lawyers and supporters of outrageous crimes, though by now there was hardly anyone left to alienate. A month or so earlier, he'd quarreled openly with Jacques Sarasin. In a single fateful confrontation, he accused his most devoted disciple and patron of having been motivated

Another one of de Loutherbourg's satires in Bienne against Cagliostro:
"Cagliostro as a fairground mountebank"

by "interest" when undertaking the mediation. No insult could have been more wounding.

Finally, Cagliostro extended this accusation to his and Seraphina's most faithful supporter in Bienne, Banneret Sigismund Wildermett. Seraphina was baffled by Cagliostro's self-destructive rage. What was the matter with her husband? Illness seemed to be turning his brain. Poor Banneret Wildermett tried to find a charitable explanation, writing to Sarasin, "One must see that all his [Cagliostro's] initial education did not form him to live in Europe; he bears the marked vestiges of his Oriental principles, so different from our own."

Seraphina knew with bleak certainty that Cagliostro had sunk the life raft that had kept them buoyant for the past seven years. Her only consolation was that nothing now bound them to this miserable country. While Cagliostro paced up and down, ranting against the boring vistas and the sanctimonious Swiss, she nagged him about her need for a sunnier climate. By July he agreed to take her to bathe her swollen joints in the therapeutic waters of Aix-les-Bains in Savoy. From there, they would test the possibility of visiting the towns and cities of Italy.

On Wednesday 23 July 1788, Cagliostro and Seraphina ate a last meal with the long-suffering Sigismund Wildermett, and at half past midnight they headed south. Reflecting the next day on this sad conclusion to the master's lofty Swiss mission, Wildermett wrote: "Abandoned to the impulses of his heart one sees a man of such goodness and *sensibilité....* I will remember him as...unique in his century, by his talents, by his heart, by his foibles, and by his troubles, compounded by his unhappiness."

Italy was not the paradise Seraphina had expected. It seemed impossible to escape the growing chorus of printed exposés or the long arm of Bourbon power. Everywhere the Cagliostros traveled during the summer of 1788 — Turin, Milan, Alessandria, Genoa, and Verona — they met a similar response. There'd be a rush of excitement from the public, eager to be treated by the famous Cagliostro; then, sooner or later, town governors would appear with an expulsion order instigated either by the Bourbons or by jealous local doctors. And there were no longer endless chains of credit from Sarasin to sustain them. Expensive hotels, well-bred carriage horses, daily changes of linen, servants in livery, fine wines and food: such necessities rapidly drained their finances.

The lovely Austrian-Italian town of Roveredo, which they reached on 24 September 1788, was no different. They were welcomed by the proprietor of the best hotel, Giuseppe Festi, and his friend Clementino Vannetti, a distinguished local scholar who aspired to write a history of the extraordinary healer. Vannetti recorded that Cagliostro was instantly besieged by "the sick in carriages, in chairs, on stretchers, until the square was filled and the crowd was crushed before the house." There were men with kidney stones and incurable cancers, women suffering from nerves or dementia, soldiers trying to rid themselves of the clap, and, above all, the elderly of both sexes desperate for the Copt's rejuvenating elixirs and aphrodisiacs.

As usual, Seraphina served as nurse and an advertisement for the beauty products. By greatly exaggerating his wife's age, Cagliostro offered her as proof of the power of his medicines. Her lovely complexion was due to the fact that "she made use of a preparation...composed of five drops.... And in mixing five drops with a very good eau de toilette, this gave to the face the whiteness of milk and the red of a ruby."

Behind her vivacious front, however, Seraphina was secretly planning to carve out a different life. She'd been pining to return to her hometown of Rome since June 1787, after a visitor secretly passed on letters from her family. Her father's tender words brought tears of nostalgia for childhood days in the vicolo delle Cripte. Her family were horrified at the report in de Morande's newspaper that Cagliostro had abandoned her in London. They prayed every day to the Madonna Santissima del Carmine, begging her to help their penitent daughter, and they asked the local priest for advice. Joseph Feliciani reported: "The advice that I give you as a devoted father is to come back to our house in uprightness. You know, my dearest daughter, that the soul is immortal and this world here is nothing. I have already taken the sureties necessary from our superiors so that, if you come to Rome, you will be protected and defended. If your husband...stops you from fulfilling the commandments of our Catholic faith, he will be punished."

As she was being shunted from one Italian town to the next, Seraphina brooded on the state of her soul and the image of her loving family. In Roveredo, she plucked up the courage to make open contact with a Catholic priest. Clementino Vannetti sensed the importance of this moment:

Now the wife of Cagliostro came with a chaplain into the church, and, kneeling, assisted at mass with devotion. And moreover, another priest, a pious man, spoke often with her of the Kingdom of God and the Church, outside of which there was no salvation; and he gave her the Acts of the Apostles and the writings of the prophets to read. And he rejoiced to see the faith and hear the good words of this woman. For, in the fervor of her spirit, she was angered with the evil sown by the so-called philosophy which flourished in France, and she rejected the modern scientific works, studying the scriptures attentively.

If Cagliostro was troubled by his wife's reversion to Catholicism, he didn't show it. As Vannetti observed, Seraphina was going out of her way at the time "to say things agreeable to her husband." When the Copt read out his latest fan letter, Seraphina, "her hair unbound and flowing about her neck, ran into the house and filled it with cries of joy. Her heart indeed was as keen as a flame, the words sprang in waves from her mouth." She hadn't been so loving in a long time.

Besides, when jealous local doctors forced them, on 11 November, to move away to the nearby town of Trent, Seraphina's recent act of piety proved handy. A letter of introduction from their landlord in Roveredo brought a warm welcome from the ruler of the town, Prince Pierre-Vigil Thun, bishop of Trent. A current of attraction crackled between Cagliostro and this bishop, recapitulating in a lower key his first encounter with Cardinal Rohan. It helped that Thun shared Rohan's fuzzy love of the occult.

It was Seraphina, though, obsessed with returning to Rome, who reminded her husband of an idea that Rohan had aired seven years earlier. The cardinal had wondered whether the church might one day recognize Cagliostro's Egyptian Freemasonry as a new kind of Catholic order. Now, with money fast running out and enemies in every town, why not try it? Seraphina excitedly pointed out that the pope was powerful enough to give them permanent protection from the Bourbons. Not even Marie-Antoinette would dare to question the authority of the Vatican. Imagine, too, the Copt as founding father of an order like the Knights of Malta!

To Cagliostro's surprise, Bishop Thun didn't rule out the idea. What a coup it would be to restore the world's most infamous Freemason to the bosom of the holy Catholic church: it would be worth a cardinal's robes. Even so, there was a way to go before Cagliostro could be rejuvenated as a Catholic, let alone made the founder of a new order. His conversations, though compelling in force and volubility, showed little knowledge of core beliefs. Thun gave the task of remedying this ignorance to a tough-minded priest, Father Ghezzi of the Church of Santa Maddalena. Ghezzi was more skeptical than his bishop that this strange Arabian Freemason would persuade the Vatican, but he admitted that Cagliostro applied himself conscientiously to the catechism and showed contrition for his sins.

The bishop was delighted. Taking the plunge, he wrote a strong letter of recommendation to the papal secretary of state in Rome. He said nothing at this point about the Copt's clerical ambitions but simply asked that the Cagliostros be given a safe-conduct to revisit the Holy City. Seraphina, he said, was a native of Rome, a good and pious woman who yearned to see her eighty-year-old father before he died. As a loyal wife, she worried that her husband's notoriety might provoke official action. For good measure, Thun added that Cagliostro had also shown a strong desire to be reconciled with the Church, having made a full confession at Thun's urging.

Seraphina didn't believe her husband's contrition but was happy if it helped get them to Rome. She said later that Cagliostro would return from his sessions with Father Ghezzi crowing at how easily he'd conned the priest. Clementino Vannetti also noticed a certain lack of reverence in the Copt's private conversations. Cagliostro loved to tell how he'd once urged a syphilitic bishop to cure himself of his pox by passing it on to a virgin. Of course, Cagliostro added wryly, the good bishop undertook this arduous cure not out of sensuality but in order to preserve himself for his flock.

On 4 April, Bishop Pierre-Vigil Thun received a reply from the cardinal secretary of state in Rome that set them all celebrating: Count Cagliostro was under no legal proscription from entering the pontifical states. No one commented that the wording of the letter was rather cryptic and ambiguous. Two weeks later Seraphina brought further good news. Her father had also written that the Cagliostros would not be troubled by the Roman Catholic authorities.

Seraphina chipped away at Cagliostro's remaining resistance with pitiful complaints about the bitter winds that swept into Trent from across the Alps. Hadn't they'd left Switzerland to escape this? How long would her health, smashed by the ordeal of the Bastille and more, survive in a town where the main waterway, the river Adige, had frozen into a solid pavement of unbreakable ice? Their money, too, was almost gone. She could see that they'd end their lives in misery — vagabonds perpetually on the run from the Bourbons, or beggars in some barbaric infidel country. Only in Rome would they find peace and security.

Reinforcement for her plea came in the middle of May when Bishop Thun disclosed the unpleasant news that he'd received a stinging letter from Joseph II, emperor of Austria and brother of Queen Marie-Antoinette, rebuking him for supporting a man known all over Europe as a swindler and an Illuminati agitator. That settled the matter; they must head for Rome. After receiving Thun's letters of recommendation to the senior Vatican cardinals, the count and countess said good-bye on 17 May 1789. Moving crisply, they paused in Vicenza long enough to pawn one of Seraphina's precious jewels in exchange for cash, then passed rapidly through Venice to reach Rome on 27 May 1789.

Home at last, yet not quite. When they arrived at the Hotel Scalinata in the piazza di Spagna, famous for welcoming foreign dignitaries, Seraphina found herself caught up in the whirl of celebrity that always attended their arrival at a new city. As usual, admirers clustered around Cagliostro. There were French expatriates who missed the sociable rituals of Parisian Freemasonry, bored Italian noblewomen like Princess Lambertini and Marquise Vivaldi who swallowed his elixirs and patter with equal zest, and even a handful of disaffected Catholic clergymen who hoped that he could advance their careers. To his delight, he was still remembered in the circles of the Knights of Malta and was immediately taken up by Knight Commander de Loras and his friend Father François-Joseph de Saint Maurice. De Loras believed that Cagliostro was a "Freemason of all grades and founder of an order that contains all secrets," who might be able to exercise some much-needed influence with that fount of clerical patronage in Paris, Cardinal Rohan.

Father François-Joseph de Saint Maurice, for his part, was already an admirer of Cagliostro. He'd been born in the Swiss valleys but educated in Paris in the Convent de Marais, where he'd dabbled in French philosophy and scientific occultism. In 1785–1786, when the Copt was a hot topic in the cafés of Palais-Royal, François-Joseph came up with a cabalistic prediction that he'd one day meet the Copt in Rome. How providential that Cagliostro should now appear, especially when François-Joseph was experiencing some hiccups in his career. For some reason a long-expected promotion to bishop had failed to materialize. Perhaps the Great Copt could help? Yes, of course he could, and Cagliostro, in return, asked François-Joseph to work as a secretary.

Meantime, Seraphina was finding her domestic situation unbearable. She'd made it back to Rome, home of her family and crucible of her faith, yet she was still tethered to Cagliostro, who, as her legal husband, forbade her from rejoining the Feliciani household. She was stuck. Cagliostro's canniness in avoiding open links with Freemasons gave no obvious grounds for denouncing him to the Vatican authorities. The grandiose plan of rejuvenating Egyptian Masonry as a new Catholic order had vanished in the face of Roman political realities—he'd been unable to wangle an appointment with the pope. Frantic with frustration, Seraphina cast around for ways of escape. One day, in desperation, she tried smearing the steps outside their hotel apartment with wet, slippery soap in the hope that her husband would tumble and break his neck. No luck: fat he might be, but Giuseppe Balsamo was still as nimble as a street fighter.

Their dwindling solvency eventually played into her hands, in two ways. First, they had to leave the luxurious Hotel Scalinata for cheaper lodgings. Seraphina recommended the much humbler Casa di Conti at the piazza Farnese, and Cagliostro had little choice but to agree. Now Seraphina was among her own people in the courts and alleys of Trastevere. Not only was their new pensione run by a family friend, the devout Filippo Conti, churchwarden of the parish of San Girolama, but it was also situated only a few hundred yards from her parents' house. Seraphina selflessly volunteered to find less expensive servants and quickly recruited two unsophisticated locals, Francesca Mazzoni as chambermaid and pious Gaetano Bossi as hairdresser. Secretly, she instructed them to act as household spies: they must lurk around corners, hide in vestibules, and take note of her husband's impieties.

The impieties came thick and fast, because Seraphina knew how to get a rise out of Cagliostro—after all, she'd lived with the man for twenty years. Nor was she the fool that many of her detractors believed. The countess could scheme with the best. Getting Cagliostro convicted of blasphemy would, she believed, enable her to get an annulment and marry again. She knew, too, that praying ostentatiously to holy pictures of the Madonna at the head of the nuptial bed would provoke her husband to fury. When he brusquely ordered the pictures taken down, she was quick to pass on her shock to the Contis, who in turn took the story to the Felicianis, and from them to the parish priest, and then to the Inquisitors.

Next, Seraphina procured handfuls of crude little holy tracts intended for the semiliterate Roman poor. When Cagliostro walked into the bedchamber, he found her poring over the pages, sounding out the superstitious phrases and crossing herself fervently. Predictably irritated, he launched into one of his displays of obscene buffoonery. He jumped around the room pointing his finger at his rear, jeering that she'd find more saints up there than in her stupid tracts. He draped ribbons on his penis and asked if she wanted to bow down to worship this holy relic. When the maid blundered into the room, she found him adorning the same organ with an eggcup while chortling, "this is the true bishop whom you must adore."

Using the same provocateur's tactic, Seraphina cajoled Cagliostro into visiting her parents, knowing that their pious conversation would inspire his usual blasphemous quips. Sure enough, he scoffed at the Felicianis for worshiping a gallows bird like Jesus and for calling the Holy Mother a virgin. He howled with laughter when a group of local children sang a naïvely worded hymn to the Madonna. And when one of the guests at the table invoked God's support to preserve Rome from the horrors of revolution, he suggested drily that God just might have more important things on his mind. These profanities were carefully noted.

Seraphina's constant complaints of poverty had a more serious effect because they finally pushed Cagliostro to consider using Freemasonry, the movement to which he was so passionately dedicated, as a way of making money. He began to spend time with a young Mason, Augustin-Louis Belle, a painter from Paris who presided over an advanced circle of students from the local French academy of art. Belle had formed a small, informal Masonic lodge, the Sincere Amis, whose members met for convivial soirées at his workshop. It was among this group, as well as a few well-to-do friends from the Hotel Scalinata, that Cagliostro now tried to set up an Egyptian Rite lodge of adoption. On this occasion, his motivation was purely mercenary: he hoped to persuade Marquise Vivaldi to serve as grand maîtresse, and, through her, to lure other wealthy socialite women to enroll.

One day in September 1789 he secretly invited a group of potential Masonic candidates to his hotel room, where he read out to them the visionary promises of his rite. They enjoyed his sonorous words, but most refused to pay fifty scudi each for an Egyptian Rite initiation diploma. He did manage to recruit two male disciples: an architect friend of the Felicianis, Carlo Antonini; and a government lawyer, Matteo Berardi. The following Sunday afternoon, the two men knelt on the floor of the hotel bedchamber to be inducted solemnly into the Egyptian Rite. Their diploma fees helped to replenish the household coffers a little. Cagliostro didn't realize, though, that Antonini was actually Seraphina's lover and spy; she hoped eventually to marry him. Neither did the Copt hear the whispering of servants in the corridor outside his room.

Cagliostro had some sense of the risk he was running, especially because a friend warned him of a rumor that his wife was trying to betray him. She'd gotten word to the Holy Office through her family that she wanted to make a deposition "for the repose of her conscience." Cagliostro took the precaution of asking Father François-Joseph to keep watch whenever he was away from the hotel, but in truth the Copt wasn't really too worried. He couldn't bring himself to believe that Seraphina didn't love him as always. He was used to her continual complaints and threats: she'd soon cheer up again, he thought, when he could get her favorite jewel out of the pawnshop.

Now it was Seraphina's turn to become alarmed, because financial pressure pushed Cagliostro toward a more radical proposal, one that threatened to blow her plan of escape to smithereens. By autumn of 1789, the papers were carrying a torrent of daily news about the rising fortunes of political reformers in Paris. This gave Cagliostro an idea. It looked as if his enemies in France were finally defeated: the king and queen were gilded prisoners, Governor de Launay had been torn to pieces by the mob who took the Bastille, and Cardinal Rohan had been invited to sit in the revolutionary National Assembly. Cagliostro and François-Joseph therefore composed a letter of petition asking permission to return to France. "Full of admiration and attachment for the French nation as well as respect for its legislators," Cagliostro longed to come back to the country of his heart, from which he'd been banned "by an arbitrary royal edict." Soon, he told Seraphina, they'd be in a country that had banished all superstition and clerical rapacity. With the help of Cardinal Rohan, the Great Copt would be returned to his former glory. He and the cardinal would use the great moral influence of Egyptian Masonry to restore harmony between a chastened king and a regenerated people.

Seraphina knew that she must act or lose everything. Her parents told her in early November that Dom Giuseppe Tosi, priest of the Church of Santa Caterina della Ruota, had at last gotten word from the Holy Office to take a formal deposition from her. But how could she reach him? Cagliostro and the Capuchin had become her jailers. She decided to deploy her most trusted weapon. One day she whispered to Father François-Joseph that she'd been nursing a secret passion for him. If it weren't for Cagliostro, they could live together; if only she could get rid of her odious husband. The aging Capuchin was electrified. From that moment, surveillance of Seraphina took on an entirely different meaning. He wafted around the pensione in a haze of steamy fantasies. By mid-November, Seraphina was accepting gifts from the besotted friar, and he was enjoying her excellent manual skills. Desperate situations require desperate remedies.

On 23 November, knowing that Cagliostro would be away for the morning, Seraphina's servants managed to sneak Dom Giuseppi Tosi into a small courtyard outside her bedchamber. Seraphina was shaking with fear. Tosi shouted questions up to her window, she gabbled back her denunciations, and an official notary scribbled down their words. Although Tosi had to leave suddenly because of Cagliostro's imminent return, the tribunal of the Holy Office declared itself satisfied with the results. They also took testimony from Seraphina's father and the two recent Masonic initiates. By Christmas, Cardinal de Zelada, the newly appointed papal secretary of state, believed they had enough evidence to act.

On 27 December, after Pope Pius VI finished celebrating mass on the auspicious Feast of Saint John, he visited de Zelada's house, where several other members of the congregation of the Holy Office were gathered. It was the Vatican's cabinet of war: de Zelada, secretary of state; Cardinal Antonelli, prefect of propaganda; Cardinal Pallota, prefect of conseil; and Cardinal Campanelli, pro-dataire. After listening to a short address from the pope, they unanimously endorsed his proposed course of action. The gravity of the occasion may be judged by the fact that Pius VI was personally involved in a matter that would normally have been left to his officials. Pius himself issued the order to the governor of Rome to seize Count Cagliostro and his secretary and to search their premises for incriminating evidence. Cagliostro was to be taken to the Roman prison of Sant'Angelo, Father François-Joseph to the Grand Convent of the Capitole.

In Trent, a shocked Bishop Thun learned later from his Roman agent that Cagliostro was deep in conversation with friends when a picket of grenadiers of the Rossi regiment stormed into the Casa di Conti to arrest him. Sensing the source of his betrayal, he picked up a loaded pistol, pointed it at Seraphina, and pulled the trigger. The gun misfired. Bellowing in rage and pain, Cagliostro was dragged away by the guards. To Seraphina's surprise, she herself was escorted to the nearby convent of Santa Apollonia. It was for her own protection, of course, and she could enjoy the endless hours of spiritual devotion that the Holy Office knew she wanted.

Seraphina's betrayal shattered Cagliostro. The jailers who watched over him in the cramped, filthy cell in Sant'Angelo and the Inquisitors who questioned him for fifteen long months, later testified to his pitiful obsession with his wife. Sometimes he called out in anguish, asking why she'd abandoned him, only to soothe himself with the conclusion that his *cara figlia* had been forced to testify against her will. Sometimes he pleaded with his warders to let her share his cell, or at least allow her to visit. Deprived of paper and pen, he couldn't pour out the tender words that he'd written in the Bastille. His interrogators noticed that he appealed constantly for her endorsement, as though she were present in the room. Tough as they were, his devotion touched them.

Seraphina was, of course, their star witness. Although the tribunal overseeing the case marshaled testimony from hordes of minor spies and informers—the prying domestic servants at both Roman hotels, the sanctimonious Feliciani family, the two recent initiates to the Egyptian Rite, the painter Augustin Belle—none of these had much substance. Silly superstitious gossip was not enough for a case of this importance. No, the Inquisitors must rely on Seraphina, the woman who'd shared his bed for twenty-one years and who alone could tear off the veil that hid his many lives. Curé Tosi was several times sent to visit her at the Convent of Santa Apollonia in order to add flesh to the skeletal details he'd collected in the Casa di Conti.

Cagliostro's strategy was simple: first, he denied any real involvement with either orthodox or heretical Freemasonry. Despite having had overtures from revolutionary Illuminati who'd been desperate to recruit him, he'd rejected their bribes. In fact, he was really a Christian who scrupulously observed the religious practices of every country he visited. He denied that Egyptian Masonry was heretical or irreligious. It was a deeply spiritual—indeed, a Roman Catholic—movement. Consider the hymns and prayers used in his services. Consider the lofty morality he always preached. He had been moved, like many Christian prophets before him, by a mysterious "internal impulse" to devote the remainder of his life to God's work of healing the sick and helping the poor. God had filled him with "beatific visions," and everyone knew that God "could shower his grace on whom he pleased, even upon sinners."

What a different story Seraphina told. Here was a man, she claimed, who shamelessly pimped for his child wife, teaching her to flash her ankles, thrust out her bosom, and sway her hips so as to seduce his victims—all the while impressing on her that she was committing no sin. Here was a man who refused to allow his wife to practice her holy faith. In private he mocked all religion and blasphemed while pretending in public to be filled with piety. He practiced black magic and made sacrilegious appropriations of the divinity of Jesus. He read irreligious books and collected satires about the king and queen of France. He was a lecher and an imposter who seduced serving-maids and washerwomen, composed drunken sermons, and bribed his tiny *pupilles* to perform at séances.

Seraphina didn't realize that this testimony was helping to dig her own grave. The more she mired her husband in filth, the more of it rubbed off on her. Though Cagliostro could not bring himself to mount counter-accusations against his darling wife, the legal advocates Constantini and Bernardini, whom the Vatican had appointed to defend him, had no such scruples. Both were able and committed lawyers. Given the highly circum-scribed scope of the legal defense within an Inquisition trial, there was no hope of winning their case, but they were determined to put up a fight. They concentrated on undermining Seraphina's credibility and character. What sort of woman would betray her doting husband after enjoying twenty-one years of luxury? If she'd been so badly mistreated, why had she not left Cagliostro earlier? What sort of woman would throw herself so enthusiastically into love affairs with Count von Howen in Courland, Prince Potemkin in Russia, and Monsieur Duplessis in France? How reli-able was this pious witness who accepted gifts from the accused Capuchin François-Joseph and masturbated him repeatedly? What were her real motives in trying to get her husband convicted of blasphemy?

In the end the two lawyers couldn't make their accusations stick, because the verdict rested with an Inquisition tribunal that had decided on Cagliostro's guilt before the trial began. But one thing did become clear as the case proceeded: Cagliostro's wife was becoming an embarrassment to the prosecutors. Though not physically present in the courtroom, she was, in a virtual sense, standing trial alongside her husband. Seraphina didn't know it, but the Holy Office decided at some point that she was both too unreliable and too important to be allowed back into the outside world.

So, when her husband heard his eventual verdict of life imprisonment, a shadow verdict was also being passed on the absent Seraphina.

Cagliostro was sentenced to see out the remainder of his days in the confines of the fortress of San Leo; Seraphina would go crazy in the cloisters of the Convent of Santa Apollonia, far from the jewels and riches she had so loved. Late at night in her austere cell, would she ever come to regret betraying the imposter who loved her?

· 7 ·

Heretic

Heretick: One who propagates his private opinions in opposition to the
catholick church.

CARDINAL DORIA TOOK CHARGE of the prisoner Giuseppe Balsamo on 20
April 1791; it would be the legate's last day of real peace. The prison coach
and armed escort party had left Rome four days earlier, traveling at night
in great secrecy so as to avoid an ambush from Balsamo's followers. When
they reached Pesaro, headquarters of the papal duchy of Urbino, Adjutant
Grilloni handed his personal instructions to Cardinal Doria, ruler of the
duchy and the administrator responsible for the prison of San Leo thirty
miles away. Grilloni's orders had come from the secretary of state in Rome,
Cardinal Francesco Saderio de Zelada.

The orders were terse and to the point. Cardinal Doria, legate of Urbino,
and Sempronio Semproni, governor of the prison of San Leo, were made
personally responsible to His Holiness Pope Pius VI for ensuring that Giuseppe Balsamo remained locked in San Leo "for the term of his natural life
without the possibility of a pardon and under strict custody." Balsamo was
forbidden to converse with anyone inside or outside the prison and was

not to be allowed any means of writing or communication with the outside world.

What Doria knew about the prisoner, aside from extensive newspaper gossip, had come from poring over an official compendium of Balsamo's life and legal trial compiled by the Vatican notary, Father Marcello, a Jesuit, and published under the pseudonym Monsignor Giovanni Barberi. Though this *Life of Giuseppe Balsamo* had only just been released by the Apostolic Chamber, a copy was rushed to Doria to prepare him for the task of superintending the man's custody. The cardinal learned from the work that it would be a grave mistake to underestimate this wily Sicilian.

On meeting the man, Doria thought it obvious that two years of incarceration in the Roman prison of Sant'Angelo had done nothing to diminish his arrogance; Balsamo acted as though the escort was in his honor and treated Grilloni with easy familiarity. Apart from his distinctively crafty eyes, the infamous Count Cagliostro was not much to look at; squat, fat, and swarthy, with a balding head. Yet Balsamo's crimes against the church were so heinous and his international fame was so great that Pius VI had kept a close watch over the entire trial and attended a number of the Inquisitor's interrogations. Furthermore, Pius would continue to monitor Balsamo's progress through the regular and detailed reports that Doria was expected to send to the secretary of state in Rome.

The legate was impressed at the thoroughness of the investigation into Balsamo: the Holy Office and Inquisition had mustered their most potent forces against the man. A panel consisting of the advocates Paradisi and Cavazzi, with Abbé Lelli as a notary, had done the arduous preliminary interrogations. They in turn reported to the special tribunal headed by de Zelada and supported by three cardinals (Antonelli, Palotta, and Campanelli), the Roman governor Rinucinni, and the Inquisition's legal expert, Roverelli. Helped by Fiscal-General Barberi (Father Marcello), who served as a special notary, this group sifted the evidence, shaped the charges, and interrogated the prisoner on particular issues.

Forty-three close interrogations over fifteen months enabled this crack team to uncover the shocking contagion of Egyptian Freemasonry as Balsamo (alias Count Cagliostro) had carried it through the length and breadth of Europe. Expert theological consultants summarized his Egyptian ritual and dogma; a team of medical experts and chemists headed by Doctor Micheli analyzed all the man's elixirs and medicines, seeking to explain his ability to heal illnesses and influence child mediums. Legal investigators were sent to the known sites of his activities in order to collect evidence of religious and civil crimes. This collective effort produced enough to charge Balsamo on three separate counts—as a member of the outlawed movement of Freemasonry; as a violator of the common laws of several countries; and, most pertinently, as a heretic with regard to the beliefs of the Roman Catholic church.

Naturally the Inquisitors had taken pains to investigate the puzzling appeal of such a vulgar and ill-educated man. They conceded that he had shown an eerie ability to debauch the human heart, particularly in those countries where morality had already been sapped by Protestant and irreligious ideas. "Count Cagliostro's" pose as an Oriental seer and his smattering of medical knowledge undoubtedly gave a veneer of authority to his ministrations and alchemical experiments. He presented himself as a benefactor of the poor, gaining a worldwide reputation for philanthropy, which—the Inquisition said—he'd used for the purpose of swindling the rich and credulous.

The Inquisitors found that the dangerously seductive secret structures of Freemasonry had provided a perfect cover for Balsamo's crimes. Emotionally charged séances had cleverly exploited the suggestibility of lonely women and overimaginative children who longed to communicate with angels and spirits. Self-confident and lucky predictions, based usually on prior intelligence or simple common sense, had given Balsamo the reputation of a seer. By appealing to "the genius of his listeners," his bombastic and obscure orations—delivered in a gabble of Sicilian, French, and Arabic—could seem profound. His ability to speak "parabolically and enigmatically" masked the true poverty of his ideas. Under a veneer of pagan spirituality, promises to rejuvenate the lecherous and shower the greedy with alchemical gold had sapped Catholic morality "front and rear."

Still more seriously, Doria learned that the Inquisitors had uncovered irrefutable evidence of Balsamo's subversive political mission. From the outset the rootless Sicilian had used Egyptian Freemasonry to undermine legitimate government and religion. He admitted that he'd joined the world conspiracy of the Bavarian Illuminati, though he pretended to have had no interest in the secret society's republican and atheistic agenda. Having been imprisoned by King Louis XVI of France as a swindler, he'd somehow corrupted the parlement of Paris enough to be acquitted, after which he'd sworn a terrible vengeance against the Bourbons. As a result—the compendium showed—Balsamo had fostered the bloodthirsty revolution now sweeping through France. He had both predicted and exulted in the fall of the Bastille, the humiliation of the king, the abolition of the clergy, and the plundering of the Catholic church.

As Doria read Barberi's compendium, he could see that the fruits of Balsamo's plots were ripening wherever the man had been. Europe teetered on the edge of a major war in a desperate effort to check France's revolutionary ideas from sweeping away the continent's thrones and altars. Poland, an early target of Cagliostro's Masonic misson, had already succumbed to revolution; so too had the Dutch Low Countries where he had also sowed his Masonic sedition. Belgium and Savoy seemed in peril from French armies, and the weakly armed papal territories would surely be next. Only Austria and Prussia seemed to be trying to hammer out a serious alliance of resistance.

Doria was not surprised, therefore, to read in the compendium that Balsamo had visited Rome in 1789 with the deliberate intention of inciting rebellion among the city's poor and in the crowded criminal sectors. Secreted among his papers, the investigators discovered a prophecy that Pius VI would be the last pope. With unbelief roaring through France, the Holy Father had naturally feared that the infection might spread to his own dominions. In December there had been reports of rebellion in the provincial municipality of Sininglia, and Cardinal Chiaramonte also reported an uprising in the Marches immediately adjoining Doria's own jurisdiction. Though it had not yet spread to the duchy of Urbino, Doria had to prepare for that possibility. The Inquisitors had found evidence of a secret correspondence between Balsamo and other Freemasons in Rome, Naples, and Venice. His Masonic lodges were said to have 180,000 members—some newspapers put the figure in the millions. He'd openly boasted to his interrogators that wealthy bankers in Switzerland, Holland, and France plied him with credit: Barberi suggested that the money was to fund a fleet of 5,000 vessels to carry invaders against the papal states.

Only the swift and vigilant action of the Roman authorities and the Inquisition's spies had so far enabled the church to forestall these uprisings. After commuting Balsamo's death sentence to life imprisonment in April 1791, the Holy Office rounded up other suspected Masons. At the same time they increased the sentries at strategic posts around Rome and passed a series of laws clamping down on popular meetings within the city. They even canceled the festivities and illuminations of the annual Easter carnival— a mob of intoxicated revelers in the city posed too serious a risk. Spies reported that Balsamo had ordered his fanatical disciples to burn down Sant'Angelo prison and release him. Rumors also reached the Vatican of Masonic plots to stage simultaneous uprisings at key papal towns.

As long as Balsamo stayed in Rome, he would be a standing incitement for attempts to escape and for invasions. A week after his sentence, therefore, the Holy Office notified Doria of its intention to move the prisoner to the remote and impregnable fortress of San Leo. Suddenly, the legate was to be in charge of the most infamous and dangerous prisoner in Europe.

Cardinal Doria was an educated and ambitious man; he knew that taking charge of Giuseppe Balsamo was the opportunity of a lifetime. From administering a small and sleepy church duchy, he'd been catapulted into intimate contact with Cardinal de Zelada, the powerful secretary of state, and, through him, with Pius VI himself. This was the stuff of dreams. Every week Doria would have to write a detailed account to Rome on the state of the prisoner. Few cardinals outside the inner sanctum of the Holy Office could ever hope for such a chance to display their tact, efficiency, and loyalty. At the same time, Doria knew that he must not on any account squander this great blessing. He congratulated himself on having made the effort to inspect the fortress of San Leo the previous year, soon after taking up the position of legate. The visit had been a truly horrifying experience at many levels, but it gave him an idea of the challenges involved in fulfilling the papal orders.

First, he had to ensure that the prison party actually reached the fortress without mishap. He warned Adjutant Grilloni that although San Leo was only some thirty miles away, it would be the most testing passage of their journey. The terrain was precipitous and vulnerable to ambush. With this in mind, Doria had already dispatched a trusted relative to scout in advance for signs of danger and to alert the garrison to prepare for their arrival.

Such precautions were necessary. Twelve miles away at Verucchio the carriage road ceased, so they had to switch to foot and horseback. The party wound along a narrow, crumbly track worn into the mountain rock, then descended, slipping and scrambling, into a wooded, boulder-strewn valley. Reaching the bottom, they were confronted with the snow-swollen Marecchia River. Wet with spray after skirting the edge, they began to climb one of the steepest goat tracks in Italy. The Apennine fang on which the tiny village of San Leo clung rose 690 meters out of the valley floor, looking over a landscape of jagged cliffs, plunging ravines, tumescent boulders, and thickets of elder. Eventually, the party reached the old stone village of San Leo; they entered through the gated portico into the main square, past the elm tree on the spot where Saint Francis had preached a famous sermon in 1213, to see the fortress looming above them. The papacy's most fearsome

state prison was built on the tip of a
further scarp of rock, sheer on all sides.
The main building narrowed to a sharp
point that cut into the sky like a dor-
sal fin throwing up ripples of cloud
around four supporting towers.

One of the most massive military
buildings in Italy, San Leo had been built
on the site of an earlier garrison for the
dukes of Urbino by a famous fifteenth-
century architect, Francesco di Giorgio
Martini. Dante evoked its awesome
appearance in the fourth canto of the
Purgatorio, and Machiavelli called it the
strongest fortress in Europe. The scars

*Iron mask used to intimidate
prisoners of San Leo*

of failed sieges ran up and down its stone sides. Yet this sort of medieval
pedigree, as Doria well knew, presented serious drawbacks in the year 1791.
Despite the seven-foot-thick walls and heavy turrets, Rome could afford
less than twenty soldiers to protect San Leo's bloated girth. Such a tiny garrison might be adequate to control the eight or so prisoners of the Inquisition inside, but it would not be able resist an invasion by an outside force,
particularly at the present time, when one of the crumbling lower sections
of the fortress needed rebuilding. These were not peaceful times.

Internal security was little better. Most of the cells had fallen into dis-
repair, making them unsafe and unsuitable for housing a prisoner of Balsamo's celebrity. Besides, the internal layout of the fortress was not designed
to keep prisoners isolated from each other, still less from their jailers. Yet
this was what the papacy demanded for the new inmate. How, for example, could Doria hide Balsamo from other prisoners at the communal daily
mass in the prison's narrow central chapel? The legate decided for the
moment to exclude him from formal worship altogether. This could be

only temporary, however, because Bishop Terzi, head of the diocese of Montefeltro, was charged by the Holy Office with looking after Balsamo's religious needs, a task he would undoubtedly exercise with all the earnestness of the pastoral clergy.

Bishop Terzi agreed for now that the prisoner would not be harmed by a period of solitary reflection "to purge his soul from that which burdens him," but Doria thought it was overzealous of the bishop to send a confessor to Balsamo on the first day. Actually, for a time it looked as if the priest might have to administer the last rites: the burly Sicilian threw such a convulsive fit on being taken into his cell that the four jailers gripping him were afraid he'd die. Having read Barberi's compendium, Doria wondered if the fit had been staged; he wrote to remind the bishop as tactfully as he could that "the prisoner is an impious and irreligious man, capable of committing all kinds of iniquities." When Doria looked over the chapel later, his quick mind came up with a solution to the problem of Balsamo's worship. The legate would ask the prison architect Baldini to build a type of *corretto*, or small booth, adjacent to the chapel. This would enable the prisoner to attend mass without being seen or heard by other members of the congregation.

Innovations of this kind were tricky because all prison expenses had to be approved by the tightfisted paymasters in Rome. Also, the governor or castellan of San Leo, Sempronio Semproni, was as old-fashioned as the fortress itself. He was loyal and conscientious, certainly—he'd looked after the prison for many years without incident—but the length of his tenure now presented something of a problem. Born and bred in the nearby hill town of Urbino, Semproni was a typical local man; kindly, parochial, and complacent, incapable of thinking about anything except petty concerns such as the marriage prospects of his daughters. And because of the severity of the weather on the exposed heights of San Leo, he returned home for three months every winter, leaving the prison in the hands of the second in command, or *tenente*, Lieutenant Pietro Gandini.

Here was a different kind of problem. This bumptious and fickle young man took delight in subverting the elderly castellan's orders. Because of his well-to-do background, Gandini could afford to accept a minimal salary. His freedom from concerns about money led to a willful refusal to bow to the authority of superiors. Doria had hinted to Rome about replacing him; but, as always, the secretary of state thought only of the cost. Lieutenant Gandini was cheap; what else was there to say? Doria decided to keep a close watch on Gandini's behavior through his personal commissary, who was under strict instructions to serve as the legate's eyes and ears in both the village and the prison.

As luck would have it, Semproni happened to be away from the prison with a cold at the crucial time of Balsamo's arrival, so Gandini delightedly took charge. Doria's letter of instruction ordered him to lock the prisoner in the "Pozzetto," or "cell of the well," famous for having housed some of the Inquisition's tougher nuts in the past. Roughly nine feet square, it had originally been gouged out of the hard mountain basalt to hold drinking water. Later, three massive isolated walls had been added, with a fourth being connected directly to a fortified tower. Its tiny triple-barred window looked directly onto a platform where two sentries kept watch round the clock. One could just glimpse the parish church of San Leo below—an edifying image for an unbeliever like Balsamo to contemplate for the rest of his days. Another advantage of the Pozzetto was that it had no door. One entered from above through a small timber-and-iron trapdoor in the roof, an arrangement that allowed food to be lowered into the interior without exposing guards to the inmate's wiles. In effect, prisoners in the Pozzetto were buried alive.

With breathtaking arrogance, Gandini instead lodged Balsamo in the Tesoro, or "treasury cell," at the opposite end of the prison. His cocky explanatory letter stated that the Pozzetto needed substantial repair work and that the Tesoro was anyway a safer place of incarceration. Since it was situated at the topmost point of the main fortress building, the Tesoro, he asserted, would isolate the prisoner completely from visual or aural contact with the outside world. Its high, tiny window, protected by four rows of heavy bars, looked into an abyss of swirling mist. As the former stronghold for the castle's treasure, its walls were nearly eleven feet thick—strong enough to defy any bombardment. Unlike the Pozzetto, it had the great advantage, too, that Balsamo wouldn't be able to hear the chatter of the soldiers, "all of whom I cannot vouch for." So confident was Gandini that he even sent away the workmen who came to repair the Pozzetto. To compound his cheek, he also wrote to de Zelada arguing against Doria's immediate and urgent request for eight more garrison troops. According to Gandini, they were unnecessary: any such addition would only "fatten those that are already fat, and put smiles on the faces of those who want to live without toil."

The legate was furious. Living in the tiny backwater of San Leo, Gandini was oblivious of the din of publicity that Balsamo's arrest had generated. Nor did he understand that their entire familiar world now stood in peril. While Gandini played silly games, the threat of democratic revolution was advancing remorselessly toward their duchy. From the moment of Balsamo's capture, the eyes of all Europe had turned to the Vatican. Every journalist clamored for news of the famous Cagliostro. Because of this international pressure, Pius had ordered that Balsamo be treated "charitably." Should the prisoner harm himself or die from ill treatment, it would reflect badly on the Holy Office. In these troubled times the Church couldn't afford to look like a medieval relic. Gandini's boast that the Tesoro resembled a "strict monastery" completely missed the point; Balsamo was supposed to be locked in a modern prison. Doria recalled the Tesoro as "a truly horrible and unhealthy cell which I do not believe would be in keeping with the wishes of the sovereign." Its walls and floor dripped with dampness; icy winds whipped through the high window; bugs and vermin swarmed over the stone floor. The prisoner's health would not long stand such conditions.

For want of an alternative, Doria couldn't immediately reverse the cocky *tenente's* decision, but he consoled himself that the more pliable castellan would soon be back. Within weeks of Semproni's return, work on the Pozzetto resumed, including the construction of a new grille on the window to prevent Balsamo from talking with the sentries. For once, too, Rome relaxed its usual parsimony. By the beginning of summer de Zelada agreed to pay the costs of the *corretto* and of transferring six more soldiers from Ancona to strengthen the garrison. During these early weeks, Doria also worked hard to eradicate several abuses of discipline that had crept into the prison regime. These included Gandini's habit of allowing prisoners to purchase their own food and even to join his family table.

Doria was encouraged that his warnings about Balsamo's "astuteness" and the need to maintain an absolute cowl of secrecy seemed to be getting through to the garrison. He drummed into the castellan that "a small oversight when dealing with the custody of this subject could produce enormous consequences," and that "no amount of diligence will ever be too much." Even the pettiest details of Balsamo's regimen had to be scrutinized. Doria suggested that the prisoner ought not, for example, to be allowed shaving implements, since he could use them to harm himself or others. He must be given no paper or writing tools of any kind—Barberi's compendium had shown that the man had a genius for publicity. Nor should Balsamo's religious pretensions be indulged; he would very likely try to seduce the mind of his confessor. A close watch should also be kept on Balsamo's relationship with the prison physician, "since the prisoner gives himself out to be a great professor in this science." Balsamo's talent for feigning illness had been evident from the moment of his arrival; now he was pretending to have problems walking in order to make the guards less vigilant.

Doria reiterated that no strangers should be permitted to enter the fortress under any pretext. This prohibition, he ordered, must be extended to the village of San Leo itself: "Remove…anyone meddlesome or anyone that does not have a reason for being there," he told Semproni. The governor was ordered also to keep a secret watch on his troublesome second in command, though Doria advised him to do this subtly so that the quarrelsome *tenente* could not complain to the Vatican.

By the beginning of the summer of 1791 Doria felt pleased with himself. He'd progressed a long way toward putting the previously slack, ramshackle administration of the prison onto a sound footing. Under his firm regimen even Balsamo seemed to be showing signs of contrition, though, as Doria told the secretary of state, "only time will tell whether his reformation is true." In his report of 12 May, the legate allowed himself a note of quiet self-congratulation: "In short...I am confident that there is no place from within the fortress where one needs fear an incident occurring, as everything has been taken care of."

Had the legate possessed the powers of prophecy attributed to Cagliostro, he might have been more circumspect. His complacency was shattered on 29 June by a curt, frosty letter from the secretary of state. News had reached Rome, through a spy, that several noblemen from the nearby town of Rimini had been allowed to enter the fortress and speak to Balsamo at length. This was completely unacceptable. Had Cardinal Doria not understood the explicit and repeated instructions of His Holiness Pius VI? The Holy Father was furious: he intended to begin a searching inquiry to discover whether this blatant contravention of his orders stemmed from political malice or neglect of duty. The legate had better get to the bottom of the matter and report back at once.

Angry and humiliated, Doria ordered his most trusted spy, Cristoforo Beni, an officer with a good knowledge of the Montefeltro parish, to insinuate himself into the village of San Leo under the pretence of investigating issues of border jurisdiction. He was to find out whether Balsamo had managed to communicate with outsiders and by what means. Above all, he must find out if these noblemen from Rimini were revolutionary sympathizers with the French. Was there a network of plotters in the village and the prison? What had these men said to Balsamo and he to them? In defending himself to de Zelada, Doria blamed the castellan and the *tenente*, both of whom he'd instructed repeatedly to allow no strangers or visitors into the fortress.

A week later Doria reported the results of his inquiry: his spy confirmed that four noblemen from Rimini had indeed been permitted to see and possibly speak to Balsamo. His spy did not think anything dangerous had been said, but Beni had uncovered a further regrettable violation. He'd discovered that the lord bishop of Penna and the bishop of Urbino had also been granted access to the prisoner—perhaps acceptably, given their spiritual eminence, but they'd been allowed to take along their families and friends as well. Any censure, Doria suggested, was of course a matter for the Vatican, but he himself would deliver a stinging rebuke to the castellan and also terminate the position of his personal commissary, who had failed to carry out orders to report "down to the minutest detail of what may happen peradventure in the City of San Leo."

On receiving the legate's dressing-down, Semproni begged forgiveness for his "rash oversight," yet his feeble excuses gave Doria an uncomfortable feeling that the elderly castellan could neither understand nor cope with his new responsibility. The man simply did not realize that Balsamo's custody had utterly transformed the significance of his fortress, San Leo. He was now holding the most notorious prisoner of the century, not the handful of blasphemers or petty criminals who'd had been under his care for the past decade. The legate wearily restated the papal orders in all their stark simplicity: "Balsamo...must never speak with anyone, nor see anyone, nor be seen by anyone." Doria suspected that in slacker times, before his appointment, a corrupt practice had grown up of allowing local busybodies to pay for the entertainment of speaking with the prisoners.

While the embers of this affair were still smouldering, a new panic flared. On 5 August the legate received a note from the governor of Macerata on the other side of the Apennines reporting that two or three Frenchmen had passed through the village of Morrovalle the previous week, apparently on their way to San Leo. Though they'd looked ordinary enough, being middle-aged, of medium build, and dressed in walking clothes, their conversation aroused suspicion. After a few drinks, they let slip that they were fervent admirers of Cagliostro, furious at his imprisonment. One had dropped a menacing hint that red wine and French tobacco would soon alleviate the count's present sufferings. Alarmed, Doria dashed off a warning to Semproni to prepare against a possible attempt by French revolutionaries to liberate Balsamo.

Semproni's response confirmed his complete disconnection from the world outside Urbino. He replied airily to Doria that if the French travelers turned up, he'd simply lock them away. This would be no problem, "since the number of people we are dealing with is small," and Balsamo was unreachable in the Tesoro. Doria, exasperated, tried to shake the man's bovine complacency by forwarding a copy of an anonymous letter he'd just received from Lombardy. It warned the legate to expect a surprise attack by French revolutionaries riding in hot air balloons, which could easily be flown over the fortress walls to land within the enclosure. After all, the informant pointed out, the French were the inventors of this amazing new form of transportation. Had it crossed Semproni's mind, Doria wrote in an acid accompanying note, that these Frenchmen whom he'd so lightly dismissed might be scouts for an aerial invasion? The legate's letter would have been stronger still had he known that Balsamo and several of his disciples regarded themselves as experts on balloon design and navigation.

Poor Semproni had scarcely heard of such things: as far as he was concerned these ideas belonged to the fantastic realm of wizardry. Still smarting from the trouble with the noblemen of Rimini, his reply teetered between incredulity and despair. "Perhaps none of these suspect persons will turn up here, but should they ever manage to put into effect the flying balloons project I would not know with which wall or with what forces to free myself from such an attempt, but because I believe it to be inexecutable I cannot therefore envisage how to defend the Fortress from hippographs or from a flying armada! I hope that the regulations governing aerostatic travel will not be published in our time." In his despair, Semproni also transformed the legate's sarcastic speculations into reality. The garrison scanned the heavens anxiously for several weeks in expectation of the aerial fleet. Rumors of the possibility also flashed around San Leo, then leaked gradually out into the wider world and reached the sensation-hungry newspapers. Before long, stories of a rescue attempt mounted by Jacobin balloonists had made their way to Paris, London, and Saint Petersburg. With Europe edging toward war, mythmakers gobbled up such morsels and digested them as facts.

Within the prison, Semproni countered the perceived threat of an aerial rescue by transferring Balsamo to the now-repaired Pozzetto. Here, deep underground, no balloonist could reach him. During the preparations for moving Balsamo, however, a search of his clothes revealed another embarrassment—the prisoner had managed to flout the papacy's total embargo on access to writing materials. Hidden in his clothes, they found a tiny almanac. With great ingenuity, he'd made a pen from mattress straw rolled on the window post into a pointed shape. "For ink," Doria reported, "he used some candle snuff which had fallen into the cell.... By mixing it with urine using his hands he turned it into ... usable ink." In the margins of the almanac was written a prophetic threat: "Pius VI in order to comply with the desires of the Queen has caused my sufferings and those of innocence. Woe betide France, Rome, and her followers."

A search of the Tesoro cell also uncovered a wooden pen hidden in the window frame, along with a piece of sharpened bone used to extract blood for writing. Now, Doria wrote to de Zelada, perhaps the castellan might realize "how prudently and circumspectly he must watch over a man who is truly astute and who demands the most scrupulous care, if one is not to be taken by surprise or deceived."

Doria decided to improve his diminished standing with the Holy Office by showing his true administrative calibre. He set out to devise and implement a comprehensive system of rules and regulations to eliminate San Leo's inefficiency and corruption. A new rational spirit would be introduced into this antiquated and costly custodial regime. First, he needed to map the shortcomings of Semproni's system by getting the castellan to answer a series of detailed written questions. On the basis of this information, Doria drafted new procedures on drill and fortifications, military discipline and pay, the duties of sentries, the surveillance and reformation of prisoners, the cost and administration of food, and a whole series of minor matters regarding logistics and victualing The new regime, he promised Rome, "will consequently allow us to live in peace and quiet."

The legate also believed that he'd discovered in one jailer, Corporal Marini, an ideal man for the job of bringing Balsamo under proper physical and mental control. Marini was something of a thug, a tough, conscientious and unshakably obedient soldier who showed a healthy dislike for the prisoners in his charge. Even Balsamo was said to be afraid of him. Doria's first step in the taming of the heretic was to have Marini conduct "frequent, unannounced, and careful inspections" of the prisoner and his cell. Until now, searches had been undertaken at the same time every morning and evening, allowing Balsamo to hide any covert tools or weapons that he was making or using. Doria was pleased at how enthusiastically Corporal Marini adopted his further recommendation that a secret peephole be bored in the trapdoor of the Pozzetto so the guards could spy on the prisoner at any time of the day or night. Marini seemed only too pleased to devote long hours to this task.

Doria's new surveillance regime soon revealed that Balsamo was both cunning and dangerous. Within ten days of the prisoner's transfer to the dark pit of the Pozzetto, Marini discovered several hiding places hollowed out of its solid stone walls. In one was found a delicate eggshell brimming with some mysterious golden-colored liquid. On 30 October Doria reported

Balsamo's latest trick: "Recently he was able violently to raise a section of the boards from his bed—something that has never been attempted by a prisoner before—in order to extract and seize a large, long wooden dowel which he hid inside a crack of the internal stone architrave used to secure the front iron railing of the cell. He then skilfully reassembled the boards so that no one would notice." Watching through the secret peephole, Marini observed Balsamo whetting the dowelling on his window post "in order to make it more sharp and piercing." A weapon of this kind in the hands of a strong man like Balsamo could be lethal.

Now Balsamo had to be transferred back to the Tesoro so that the bed boards in the Pozzetto could be replaced with a new, nailless wooden structure bricked into the wall. The hollowed-out hiding places also had to be filled with fresh stone and lime. All this cost money, and at the same time Doria was receiving weekly lectures from Rome for spending too much on the prison administration. He couldn't win. Irritated, the legate ordered the builder to cement into the Pozzetto wall beside the bed a set of heavy shackles and chains. Despite an earlier prohibition on such barbaric punishments, Rome agreed that the castellan could now be allowed to chain up the heretic should he continue plotting to escape. Doria hoped that the threat alone might be enough "to rouse fear in him and make him submissive."

Within two months of Balsamo's return to the renovated Pozzetto with heavy iron chains and shackles, his jailers reported the discovery of another homemade screw spike, hidden in the privy duct. Again, it had been honed to a disturbingly sharp point. Of course Doria immediately received another rebuke from Rome for failing to put a stop to such threats. At this point the cardinal lost his temper. He sent a series of angry letters to the castellan ordering the guards to intensify their observation and double the frequency of searches. A builder was told to comb every inch of the cell and, if necessary, to remove the plaster and check behind it. Everything should be searched: Balsamo's body, bed, shoes, clothes, and privy; the grille; the iron bars; everything. Doria ordered that consideration be given to sealing off the privy altogether, or at least blocking in the duct with a huge piece of elm wood.

Balsamo had gotten under the cardinal's skin, there's no doubt about that. Doria's letters to Semproni began to take on an obsessive tone. Each repeated the same mantra: the garrison must not on any account underestimate the "refined malice, perfidy, and astuteness" of this detestable prisoner.

In blacker moments, Doria suspected that Balsamo was waging a deliberate psychological war against the jailers, the administration in Urbino, and the Holy Father himself. From the outset, Balsamo had adopted the pretense of being a devout Catholic who'd fallen into heretical belief out of conscientiousness. As a former novice, Balsamo knew full well that the church would feel obliged to persuade him of his spiritual errors and lead him toward sincere contrition, particularly since he declared himself an official "schismatic." Of course, Doria conceded, spiritual reformation was a laudable aim in theory, but Balsamo was simply using the church's compassion to create trouble. The bishop of Montefeltro and his parish priest, Canon Tardioli, consistently underestimated the depravity of the man. Sensing their naïveté, Balsamo had concocted a series of ingenious strategies to create disruptions and get publicity.

As soon as he arrived at San Leo, Balsamo began presenting a theater of penitence by pretending the most abject contrition in order to make extravagant demands on both his confessor and the castellan. Doria warned Semproni of these likely tactics: "Being an astute man he may contrive to be confessed often and in this way be given the means to leave his cell frequently." While Balsamo's spiritual pastors might be delighted that the prisoner was fulfilling his spiritual duty and undertaking a lengthy general confession, Doria could only caution them that "we should not be too quick to believe this; we are dealing with a very astute man, who knows all too well how to perform in order to surprise and deceive."

And perform he did. Semproni became increasingly rattled by the prisoner's endless demands for new props to enhance his spiritual mortifications. He claimed to need a special wooden prayer stool to protect his knees from the damp. He requested that a large wooden crucifix be substituted for the insulting papier-mâché crucifix provided. He disliked the image of Christ's face on his crucifix and wanted it changed so that he could reach deeper levels of penitence. "Pay less attention to the whims of this man," Doria snapped to the castellan; a blind man could see that this affectation of piety was designed to put them off guard.

Balsamo next began to undertake holy fasts. The man who had begun by gobbling down so much chocolate, pigeon, and red wine that Rome ordered a reduction in his food allowance was now subsisting on nothing but water. The jailers wondered cynically whether his real aim was to lose some weight in preparation for an escape (perhaps then he'd be light enough for the balloon to get airborne). His confessors, on the other hand, believed that this was a genuine mark of contrition on the part of someone intensely engaged with Catholic spirituality. Doria couldn't fathom Balsamo's real motive for fasting but warned that they could expect only "true hypocrisy" from the man: "The more he affects devotion, the more one needs to be careful." As the prisoner began to shed pounds, Bishop Terzi advanced a disturbing suggestion that he might be starving himself to death.

Worried by this suggestion, Doria decided to summon expert pastoral help from a Dominican lector and theologian, Friar Bussi. At first the results seemed promising: Balsamo made a healthy confession and renounced his hunger strike. But a week later, an irritable letter from Rome notified Doria that the prisoner had managed to hand Bussi a letter of complaint about the brutal Marini, written with improvised ink on a piece of cloth torn from the lining of his coat. Bussi, of course, blabbed the story all over the Vatican. Doria's angry orders to increase surveillance through the peephole and again double of the number of cell searches then produced a complaint of excessive harshness from Balsamo's spiritual mentors. Doria felt himself in an impossible situation. He now had to grit his teeth and remind the jailers that, despite the prisoner's wickedness, "no amount of diligence is be omitted on our part in trying to procure his reformation."

One of Balsamo's most effective modes of warfare was to paint Masonic and biblical emblems and prophecies on the walls of his cell — the Sicilian had, it seems, returned to one of his earliest talents. Most of these murals were smeared on the plaster with a paint he created by blending his urine with rust from the bars of the cell. Sometimes he also used excrement. He made a brush by ripping away pieces of wood from his bed with his teeth. To this handle he somehow sewed a straw and cotton tip. He could use his improvised brush with uncanny speed; on one occasion Marini's spyhole was unattended for only an hour, during which time Balsamo managed to fill an entire wall with "loathsome drawings."

To Doria, it was clear that the prisoner also delighted in tormenting his confessors with theological conundrums. They, however, persisted in taking the man seriously. Marini reported that he'd overheard Balsamo crowing to himself over the church's legal and moral dilemmas. Balsamo asserted that the Vatican should either punish him with death if it believed him a heretic or release him as a true Catholic. The wretch took malicious pleasure in wasting time by undertaking enormously lengthy general confessions. Just before absolution, however, he would suddenly blast the hopes of his confessors by announcing that he actually doubted the Holy Father and the tenets of the Catholic church. Even the pious and patient Canon Tardioli began to wonder whether Balsamo might be an "obstinate misbeliever." At the same time Balsamo harassed the bishop of Montefeltro with complaints that his soul was being neglected because the local priest had proved intellectually inadequate to the task of instructing him. In fact, he demanded two properly qualified new confessors.

Toward the beginning of 1793, Balsamo found a new strategy. He began to create currents of unease among the jailers by making prophecies; he proclaimed that he had access to divine secrets vital to the survival of the papacy and the whole world. The shocking execution of the French king and queen and the imminence of a European war created disquiet in the garrison. Stories circulated of Balsamo's uncanny power of predicting the future. Even Doria felt obliged to hear what the prisoner was prophesying. When the man's words were transcribed, however, it was difficult to tell whether the enigmatic phrases, unintelligible symbols, and rambling stories came from a deranged or a satirical mind.

Doria assured de Zelada that he personally "was too much convinced of [Balsamo's] mockery to be fooled" but nevertheless relayed the prophecies to Rome. One prediction of November 1793 was that the pope would be assassinated in his chamber by a woman dressed as a man. Doria dismissed the story as malicious rubbish but again sent it to Rome. No one could be sure whether the prisoner was raving or mocking. Though the vision was ridiculous, it had disturbing elements. Was this particular prediction inspired by Charlotte Corday's recent assassination of the French revolutionary Marat? And, if so, how did Balsamo know about it? He always seemed to have disconcertingly accurate knowledge of the progress of the French Revolution. Given his tomblike isolation, where did he learn, for example, that his enemies the Bourbons had been executed and that revolution was being carried by France's revolutionary armies all over Europe?

Doria was afraid that Balsamo obtained knowledge by corrupting his guards. Hadn't Barberi warned that the man had an uncanny ability to exploit human susceptibilities? Doria suspected that Balsamo used this genius to sow dissension among his custodians. Despite the man's regime of isolation, he somehow managed to inflame a smoldering animosity between Corporal Marini and Lieutenant Gandini. During the winter of 1791–1792 this nagging discord flared into several open quarrels, with debilitating effects on the morale of the garrison. As always, such problems found their way to Rome. At the end of January 1792, de Zelada wrote in an annoyed tone to say that he'd been sent a petition from the prisoners of San Leo complaining of Corporal Marini's brutality. What was going on? Hadn't the pontiff himself prohibited such behavior? Doria tried to assure the secretary of state that this petition was "nothing more than the result of the prisoner Balsamo's deceitfulness." Though he couldn't prove it, the legate was certain that Balsamo had incited Gandini to help the prisoners organize their campaign against the hated corporal.

Further concerns about Gandini's links with Balsamo surfaced in July 1793 when Doria discovered that the *tenente* had been secretly allowing his wife and child to see the prisoner during mass. Here was another "significant transgression." The legate ranted to Semproni that "out of three officers not one was able to comply exactly with the sovereign orders." Two months later, he had to ask his new commissary to find out how Rome had received yet another petition against Marini, "even worse than the last." So damaging was this internal quarrel to the efficient running of the fortress that Doria wrote early in the new year begging the secretary of state to separate the two custodians by transferring Gandini to the new papal army: "in this way the disorder between them would cease and Your Eminence would no longer need to be troubled with this affair." The request was ignored.

About the same time, Balsamo began to adopt the strategy of appearing mad. Most of the jailers believed his mania to be genuine, but Doria thought he knew better. The prison physician also believed that the man's mind was disintegrating—perhaps he was suffering from the long-term effects of syphilis, supposedly acquired in Spain decades before. But Balsamo would not have been the first prisoner driven insane by conditions in the black stinking hole of the Pozzetto. In early 1792 Balsamo expressed a deluded belief that his beloved Seraphina was locked in an adjacent cell, and the following year he began calling out to her piteously and begging the guards to treat her gently. Do not harm his little countess, his *cara figlia*; she was as delicate and loving as a flower. He didn't know that Seraphina was herself rumored to have gone mad in a Roman convent.

From time to time Balsamo would suddenly howl like a tortured animal. He would run to the window of the Pozzetto and shake the bars, or he would gnash his teeth and swear at the sentries. He asked to be called "Giuseppe the sinner" and begged to be supplied with the hair shirt of a penitent monk. Doria didn't for a moment believe Balsamo was mad; this was just another example of the man's inveterate cunning. Actually the legate was beginning to sound a little mad himself. His obsession with this prisoner was obvious. And mad or not, Balsamo became increasingly difficult to control. One day he threw a chamber pot full of urine at the heads of jailers; he bit Marini; he screamed in rage when given a dirty napkin; and he kept stinking fish in his cell to offend the nostrils of his confessors.

To shut the man up, Doria had to adopt the primitive methods of punishment used by Inquisitors of former times. He ordered Semproni to chain Balsamo to the wall beside his bed and, if this didn't work, to beat the man. The castellan shrugged and obliged. He was soon reporting that one jailer had slapped the prisoner, to good effect, and on another occasion a group of guards thrashed Balsamo all over the head and the body with their batons. For a time this shocked him into better behavior, but it didn't take the wretch long to come up with an effective countermove. Semproni reported that when threatened with a beating, Balsamo now ran to the window of the Pozzetto and screamed through the bars that he was being tortured and murdered. The extraordinary power of his voice created a stir in the village. The guards had to transfer him back to the Tesoro where his terrible screams could not be heard.

Meanwhile Doria grew increasingly depressed as the French revolutionary armies advanced successfully against Prussia, Austria, and Italy. In September 1792 the revolutionaries' cannonades had blown the Prussians to pieces at Valmy. Savoy had recently been taken, and now northern Italy was threatened. God knows, Doria had done his best to strengthen the fortress against the threat of external attack. He'd persuaded the miserly secretary of state to fund the building and erection of a third reinforced gate at the entry into the village and to pay for half a dozen extra soldiers to be transferred from Ancona. But these soldiers had complained about the bitter winter cold and demanded to return home. Much to the annoyance of Rome, local soldiers had then demanded higher pay because of the extra demands imposed by Doria's system of surveillance. The villagers of San Leo also petitioned Rome against the restrictions imposed by the legate's introduction of tighter controls on the use of gate keys.

Doria knew that no gates would be strong enough to halt the plagues of democracy and war spreading through Europe. By the end of his second year in charge of Balsamo, he was grappling with the intractable problem of improving the fortress's shaky defenses. San Leo might look formidable, but an inventory of its stocks revealed woeful shortfalls in materials, soldiers, artillery, and ammunition. As always, Doria's impulse to institute reforms collided with two problems: the Vatican's parsimony and the garrison's apathy. Despite his constant reiteration that the presence of Balsamo had turned San Leo into a highly likely target for an attack by the French, no one seemed to care. By March 1793, Doria was so worried about the vulnerability of the fort that he pleaded with de Zelada to finance the appointment of a veteran officer who could train the garrison troops, even if this meant replacing the trusted Corporal Marini.

Providence itself seemed to work against him. On 17 June 1793, at 11 P.M., a huge slab of rock sheared away from the mountainside, carrying off two of the bulwarks of the fortress and part of the barrage. Most of the newer buildings were now dangerously exposed. Balsamo of course claimed credit for this divine intervention. One could be pardoned for thinking that the devil was conspiring to help his favorite son.

As the revolutionary armies swept closer, Cardinal Doria grew tired. The tone of his letters oscillated between hysteria and hopelessness at the litany of petty problems that San Leo managed to generate. Whether his comprehensive plan for the rational management of the prison ever received papal approval remains unclear; it hadn't by the end of 1792 because, he observed mournfully, "the Holy Father is engaged in many more serious matters pertaining to the government." Anyway, what was the point? Every effort to instill better order and discipline was subverted by the cunning of Balsamo. Whether or not by his own choice, Cardinal Doria abruptly left the position of Urbino's legate sometime in the early spring of 1794. He was replaced by a professional diplomat.

With Doria's departure, both the papacy and the prison tacitly conceded defeat in its attempts to control and reform Giuseppe Balsamo. On 27 May 1794 the new administrator wrote to Semproni that he was "truly saddened to hear...that the prisoner Balsamo is stubbornly keeping to his wrongful ways and...displaying even more insolence than usual." One after the other of Doria's disciplinary innovations was abandoned as too difficult, too costly, or simply ineffective. By this time even Balsamo's confessor was refusing to administer the sacraments to him because of his impiety. Prison guards found that beatings no longer quieted the man's "fierce rages" or stopped him from screaming to the residents of San Leo with "scandalous frenzy." They gave up trying to stop him from painting on his cell walls, even though the resulting images reflected "brazen misbeliefs" and "great irreligiousness." Whether he was mad or malicious, incorrigible or insane, no one could tell. One thing was certain: Balsamo's antipapal prophecies grew increasingly apocalyptic as the French armies advanced through Europe.

In a sense Balsamo's final victory against the Vatican occurred on 26 August 1795, when he suffered a sudden stroke early in the morning. Priests were rushed to his side to administer the last rites so as to save his benighted soul, but "he continually refused their salutary exhortations and was not willing to be confessed." At four in the afternoon he suffered another stroke. Giuseppe Balsamo died later that evening, "an impenitent man." By then his appearance had grown as feral as his mind. His fifty-two-year-old body was horribly emaciated, his clothes were filthy, and his beard was shaggy. Even in death his corpse reflected the failure of Doria's rational project; it looked like that of a wild man lying in an underground cave amid his own filth. Governor Semproni was so suspicious of the prisoner's tricks that he held a lighted rush against the soles of the man's feet to make certain he wasn't feigning death.

As an unrepentant heretic, Balsamo was buried in an unmarked pit at the extreme western end of the *fortezza*, between two sentry redoubts. He might have been proud to hear himself described in the formal report of the archpriest of San Leo as follows, "A heretic famous for his wicked ways, after having spread through various countries in Europe the impious doctrines of Egyptian Freemasonry, to which he made by subtle deception an infinite number of converts, fell into many vicissitudes, from which he escaped without injury thanks to his shrewdness and ability."

The revolutionary armies that Doria had so long feared finally reached the *fortezza* of San Leo in 1797. One of General Dombrowski's first actions was to demand the whereabouts of the famous Count Cagliostro. According to legend the officers of the Polish legion, on learning of the prisoner's death, ordered the remains to be dug up. Picking up his whitened skull, they filled it with wine and toasted his memory.

Immortal

Immortal: 1. Exempt from death; never to die. 2. Never ending; perpetual.

As CAGLIOSTRO PREDICTED, death made him immortal. He'd often told his disciples that they shouldn't worry on hearing of his death; it meant that he'd reached the highest grade of Egyptian Freemasonry to become one of twelve immortals ruling over mankind's destiny for the rest of time. Like the biblical prophet Elijah, he'd be carried to heaven on a chariot borne by angels; or, like the phoenix, he'd rise from the ashes of extinction.

Some disciples took the prophecy literally; the infinitely forgiving Jacques Sarasin was one of these. Four years before Cagliostro's death, Sarasin had refused to finance a proposed rescue attempt, saying that Cagliostro wouldn't have wanted it: like Christ, the master had allowed himself to be imprisoned so as to convert his enemies. Simultaneously, a completely different story also circulated among disciples—that Cagliostro had strangled his confessor and escaped from San Leo dressed in the priest's clothes. Cagliostro lived and would soon reappear; the man buried at San Leo was said to be a substitute because the Inquisition dared not admit that its nemesis was free.

But the ragged, emaciated body in the unmarked grave in San Leo was indisputably Giuseppe Balsamo's—Castellan Semproni made sure of that.

So it wasn't that Cagliostro actually cheated death, but that death anointed him as a symbol of even greater stature and influence. Freed from the picayune details of his life, he floated across time and space as freely as he'd predicted. Typically, he now generated at least as many images in his culture as he had generated during his life. Casanova, Elisa, Catherine, Jeanne, Théveneau, Seraphina, and Doria each knew a different Cagliostro, and his later mythical forms were equally protean and contradictory. Even after death Cagliostro couldn't seem to escape the same obsessive question— what was he really? A spiritual seeker, a mystic and healer guided by a deeper purpose? Or a charlatan, an opportunist, a con man working on Europe's grand stages?

The apotheosis of Count Cagliostro

One legend about Cagliostro that proved particularly enduring was the idea that he led a Masonic conspiracy against Europe's old order. Cagliostro never was a revolutionary, but many will see him no other way. The seeds of this theory were sown by Elisa von der Recke's published *Report* of 1786. Disillusionment with her guru led her to suspect that he was a crypto-Jesuit trying to use Strict Observance Freemasonry to undermine Protestant religion. Her many tours as a celebrity and as an authority on Cagliostro enabled her to spread the theory all over Europe. The outbreak of the French Revolution further strengthened her case. Even those, like the prince of Schleswig, who had a soft spot for Cagliostro were inclined to concede that he "started the French Revolution and it was to be expected that Frenchmen would free him."

One of Elisa's greatest admirers, Empress Catherine of Russia, also fostered this conspiracy theory. In 1795, she invited Elisa to Saint Petersburg to reward her for her activities against Cagliostro—by this time Catherine had already banned Freemasonry in Russia. Looking back on Cagliostro's tour to eastern Europe twenty years earlier, the two women agreed on his motives: "Cagliostro was an envoy of the order of Jesuits which, during this time, was banned and which instigated the French Revolution by circulating insanity and error by the means of secret societies in several countries."

Other enemies of Cagliostro enriched this black legend. No single source was more important than Giovanni Barberi's compendium, which claimed that Cagliostro had confessed to joining the leadership of the revolutionary and atheistical German Illuminati, who also operated under the cover of Masonry. Several Catholic victims of the French Revolution added their support. In 1792, a French Catholic priest, Abbé Augustin Barruel, managed to escape to London, where he would spend the next five years assembling a massive dossier of lies, half-truths, and errors which he published as a multivolume opus, *Memoirs of the History of Jacobinism*.

Apart from his mass of documentation, Barruel's original contribution to the conspiracy theory was to trace its origin back to the suppression of the Knights Templars in 1312. He claimed that a remnant of the Templars had formed a Masonic underground in Scotland dedicated to undying vengeance against the church. Count Cagliostro, Barruel believed, had inherited their mission. His Knights Templar lodges linked up with the anticlerical salons of French philosophers and the republican cells of German Illuminati. Out of this toxic blend came a plan of world terror, whose first success was the French Revolution.

This version of the Masonic conspiracy became a staple, imitated and extended all over the western world. The idea also migrated into literature, art, and culture through a myriad of agents, carrying Cagliostro's mythic role along with it. Over the centuries, the idea of a multitentacled world conspiracy of secret societies—whether of Templars or Illuminati—took hold in the European imagination. In his best-selling novel of 1988, *Foucault's Pendulum*, the Italian philosopher and social critic Umberto Eco suggests that those with a paranoid cast of mind were intoxicated with the idea of a secret conspiracy irrespective of any content or purpose. The secret was that there was no secret; there was merely the thrill of involvement in a vague and mysterious vanguard unknown to others. Elsewhere in his writings, Eco ascribes the classic rendition of the conspiracy to the nineteenth-century French novelist Alexandre Dumas, whose hugely selling *Joseph Balsamo*, a rollicking ten-volume historical romance of 1846–1847, brought worldwide notoriety to Cagliostro's supposed plot.

In Dumas's version, the seven European chiefs of the Illuminati gather secretly in 1788, under the leadership of Cagliostro, to hear a progress report on the conspiracy. They are from Switzerland, Sweden, America, Russia, Spain and Italy, and they include the future French revolutionary Marat. Cagliostro himself represents the Masonic hubs of Germany, France, and Holland. Each of the leaders wears a signet ring indicating high initiation. Their immediate target is to eradicate the French monarchy, but their ultimate aim is world revolution. According to Dumas, Balsamo makes them swear a chilling terrorist oath:

In the name of the crucified Son, I swear to break all the bonds of nation which unite me to father, mother, brother, sister, wife, relation, friend, mistress, king, benefactor, and to any being whatsoever to whom I have promised faith, obedience, gratitude or service.

I will honour poison, steel, and fire as means of ridding the world, by death or idiocy, of the enemies of truth and liberty.

I subscribe to the law of silence. I consent to die, as if struck by lightning, on the day when I shall have merited this punishment, and I await, without murmuring, the knife which will reach me in whatsoever part of the world I may be.

Dumas's plot has continued into our time, first as the novel, which is still read today, then as a hugely popular play of 1878 by Alexandre Dumas *fils*, and finally in the cinema. The cinema seems particularly apt, given that Cagliostro and de Loutherbourg discussed propagating Egyptian Masonry with the help of magic lanterns and clockwork models, commonly regarded as a prototype of cinema. Cagliostro would have loved the idea of attaining immortality on celluloid. The best-known, though not the best, cinematic version of Dumas was Gregory Ratoff's *Black Magic* of 1949, in which a silly story gains a charge of energy from the mesmerizing performance of Orson Welles as Cagliostro. François Rivière's *Giuseppe Balsamo* (French and Italian versions, 1979 and 1980) was closer to Dumas's story and vividly evoked the Masonic conspiracy, which is shown as failing only because the French king dies hours before the revolution was to be launched.

It would be unfair to blame Cagliostro for the actions of mythologizers, but Umberto Eco has shown that the idea of a Masonic conspiracy has borne some terrible fruit. During the early twentieth century, Jews rather than Masons became the prime target. By the time it reached a bitter young man in Vienna called Adolf Hitler, the idea had taken a new and monstrous shape. Templars, Illuminati, and Egyptian Masons had given way to the Protocols of Zion, the Twelve Tribes of Israel, and the world conspiracy of Judaism. Whatever one may think of Cagliostro, it is devastating to think that he was in some way a conduit of the Holocaust.

As we might expect, Cagliostro's legendary Masonic conspiracy was also embraced by some with political views opposite to those of Abbé Barruel or Adolf Hitler. To some admirers, Cagliostro became the hero of a magical underground. Locked in his miserable Tesoro cell, Cagliostro didn't know that propaganda by the Roman Inquisition was generating a surge of popular sympathy for him. He would have loved the irony of this. While

Cardinal Doria was frantically trying to stop his incorrigible prisoner from communicating with any living soul, the Holy Office was spreading the news of his martyrdom across the length and breadth of Europe. When cheap editions of Barberi's compendium circulated in several European languages in 1791, they triggered a heated debate between Catholics and Masons. One Mason, Wolfgang Amadeus Mozart, gleefully inserted Cagliostro into his last opera, *The Magic Flute*, as the Egyptian Mason Sarastro.

In Britain, another maverick spirit was also turning Cagliostro into art. To the British engraver-poet William Blake, who'd been nurtured on truculent Puritan traditions of liberty, Cagliostro's incarceration in San Leo was a triumph for the reactionary scarlet whore of popery. Blake's great radical prophetic poem of 1791, *The French Revolution*, presented Cagliostro as a figure of countercultural resistance:

> ... and the den named horror held a man
> Chained hand and foot, round his neck an iron band bound to the
>   impregnable wall, in soul was the serpent coil'd round in his heart,
>   his from the light, as in a cleft rock:
> And the man was confin'd for writing prophetic.

Barberi's tract was also read in the more fashionable artistic circles frequented by Philippe de Loutherbourg and his cronies. Whatever they thought about Cagliostro's character, liberal-minded Masons couldn't help sympathizing with his plight as a victim of persecution by the Catholic church. The private libraries of de Loutherbourg and his friend the mystical artist Richard Cosway bristle with pro-Illuminati, anti-Bastille, and anti-Inquisition works. Nothing better symbolizes the shift in Cagliostro's status from villain to martyr than the appearance at the turn of the century of one de Loutherbourg's most haunting watercolors—a chained madman lies on the floor of a hard stone cell staring out at us with tormented eyes.

Cagliostro as a mad prisoner in chains

De Loutherbourg also made a deeper contribution to immortalizing Cagliostro by eventually adopting his spiritual mission. Despite their quarrel, the painter returned to England from Switzerland in 1788 a changed man. He caused an immediate sensation by announcing that he'd given up painting to open a healing clinic for the poor. At their house in Hammersmith, Philippe and Lucy de Loutherbourg treated several thousand ordinary Londoners using a combination like Cagliostro's of diet, health tonics (the famous mineral water?), stroking of the hands, and divine spiritual assistance. Eventually the crowds grew so great that rioting broke out and the de Loutherbourgs had to leave for the countryside. The painter was shocked to find himself abused in prints, newspapers, and debating clubs. Disconcerted at being thought a subversive in the first year of the French Revolution, de Loutherbourg closed the clinic and resumed painting.

Instead of abandoning Cagliostro's spiritual mission, however, de Louther-bourg now turned it into Romantic art. Over the next decade Cagliostro's occult themes and theories underpinned his work as an artist. One contemporary claimed that de Loutherbourg and his friends continued in private to make "voluntary excursions into the regions of the preternatural... whose ideas are like a starry night with the clouds driven rapidly across." The same pencil that had lampooned Cagliostro as a fairground mountebank now sketched his apotheosis as an Egyptian Freemason. In 1791 de Louther-bourg began work on what became more than a hundred prophetic illustrations. One of the first in the series celebrated Cagliostro's ascent into immortality. "The Ascent of Elijah" showed a figure like Cagliostro being carried up into heaven in a chariot—Masonic eyes flashing from hubs of the wheels—while an Egyptian spirit with the face of an eagle, ox and man looks on. This Masonic ascension marks an important moment in the evolution of the myth about Cagliostro: he had become an emblem of magical creativity in an unenchanted world.

Cagliostro also became entangled with another fertile myth. By the 1780s, he was hinting at being "the man of 1,400 years, the Wandering Jew," a world traveler, a witness of history, and an immortal alchemist. Cagliostro sensed the artistic longings embodied in the story of the immortal Jew, and the opportunities for appropriation that it offered. By superimposing himself on the legend of Ahasuerus, he breathed life into a textual phantom. Scores of poets, novelists, and musicians all over Europe continued this process long after Cagliostro's louse-bitten body had been dumped into the dirt at the San Leo. As the myth of the Wandering Jew was lifted up by the wave of cultural and aesthetic romanticism sweeping through Europe, Cagliostro, as nimble in death as he'd been in life, rode it all the way to the beach. Before long, a new generation of popular writers like Eugène Sue in France, Edward Bulwer-Lytton in Britain, and Johannes von Guenther in Germany had merged Cagliostro and the Wandering Jew into a fashionably moody and anguished rebel, tilting against oppressions of the spirit.

Others carried the Copt into new realms of literature, art, and music. Most artists, though, remained uncertain how finally to judge him—whether as a divine or diabolical symbol. Friedrich von Schiller's mysterious unfinished novella *The Ghost Seer* (1789), Bulwer-Lytton's Masonic novel *Zanoni* (1842), E. Dupaty's ballet *Cagliostro or the Magnetizer* (1851), and Johann Strauss's operetta *Cagliostro in Vienna* (1875) all showed the same uncertainty about whether to depict him as a rogue or a genius.

For some, the contradictions between Giuseppe Balsamo the crook and Alessandro Cagliostro, the saintly healer, were simply too much to accommodate. One solution was to declare them separate people. A Masonic writer, W. R. H. Trowbridge, put forward the idea that Alessandro Cagliostro, a nobleman of Spanish origin, was not the same man as his evil doppelgänger, the Palermitan street thug Giuseppe Balsamo. Trowbridge's readable biography of 1910, *Cagliostro: The Splendour and Misery of a Master of Magic*, has rarely been out of print, and there is no doubt that the idea of twin Cagliostros—one good and one evil—still retains its power.

This idea has had a more recent expounder in the brilliant Italian artist, biographer, and filmmaker Piero Carpi. His famous comic books—*Cagliostro* of 1967 and *Il maestro sconosciuto* (*The Unknown Master*) of 1971—stand out among a plethora of such comics. Carpi's biography *Cagliostro il taumaturgo* (*Cagliostro the Thaumaturge*) was the inspiration for a film with the same theme. Simply called *Cagliostro*, it was produced in 1974 for Twentieth Century Fox with a screenplay by Carpi: "Mistaken for the swindler, Giuseppe Balsamo, Cagliostro has up to now suffered from historical falsehoods and a conspiracy of silence.... Piero Carpi, famous expert on magic and mysteriosophy, lifts the veil covering the mystery of Cagliostro, presenting...the face of a fascinating man, human and divine, son and father of his own legend."

Elisa von der Recke, who did so much to make Cagliostro an immortal, died in April 1833, the same year that the Scottish historian Thomas Carlyle wrote his brilliant polemical essay, "Count Cagliostro: In Two Flights," which moved Cagliostro's myth in an important new direction. Drawing on inchoate ideas that Carlye found in writings of Cagliostro's chief enemies—especially Elisa, Casanova, Catherine, Théveneau, Barberi, and Jeanne—he fashioned Cagliostro into a historical emblem. Carlyle sensed that the magician's thunderous impact on his times made him a symbol and symptom of his age. Cagliostro's celebrity became a "sign of the times" and an expression of deeper historical forces.

The germ of this notion had been present in the last sad pages of Casanova's *Soliloque d'un penseur* (1786), when he ruminated on the mystery of Cagliostro's success and that of other "dismal" quacks. He was puzzled by the change in the tide of history that would celebrate Cagliostro as a genius but turn Casanova into a nobody. Although he was separated from Cagliostro by less than a generation, Casanova sensed that his rival had benefited from a change in taste. Casanova took pride in being an enlightened adventurer of the old regime—urbane, polished, witty, and genteel; Cagliostro belonged to the darker chaos of plebeian revolution and Romantic irrationalism. The world seemed to love Cagliostro precisely because he was a savage. "This is a man," wrote Casanova, "whose partisans think him wise because when he speaks he seems ignorant. This is a man who is persuasive because he is master of no language. This is a man whom people understand because he never explains himself.... This is a man whom people believe noble because he is gross in his discourse and manners. This is a man whom they believe sincere because he has all the appearances of being a liar." The old count died in 1797, soon after receiving a food parcel from Elisa von der Recke. To the last, he remained convinced that something terrible had happened to the world that he'd known, but he couldn't describe it exactly.

Thomas Carlyle was confident that he could. His essay on Cagliostro's "two flights" set out to explore the same central mystery as Casanova's *Soliloque*—the mystery of Cagliostro's success. How, Carlyle asked, in an age of supposed enlightenment, had someone without intellect, beauty, or charm managed to dupe or dazzle a great percentage of the mightiest nobles, the greatest churchmen, and the foremost thinkers of Europe? True, Carlyle conceded, Cagliostro had shown some exceptional negative abilities: "cunning in supreme degree," "brazen impudence," "a genius for deception," "pig-like defensive ferocity," "a vulpine astucity." He was also an artist in the medium of wonder, someone who could ignite torpid imaginations. In short, Cagliostro was a perfect expression of his type—"a quack genius."

As with all geniuses, Carlyle argued, what lifted Cagliostro from ordinariness was that he gave supreme voice to the needs and longings of his times. Balsamo had burst into Europe at a moment of extreme social decadence and the "utmost decay of moral principle." National dishonesty had brought Europe to the verge of an elemental explosion: "what boundless materials of deceptability, especially what prurient brute-mindedness exist over Europe in this the most deceivable of modern ages."

Cagliostro's was an age not of reason, Carlyle insisted, but of quackery and imposture—and the Sicilian was its sublime archetype: "... Quack of Quacks! A most portentous face of scoundrelism: a fat, snub, abominable face; dew-lapped, flat-nosed, greasy, full of greediness, sensuality, oxlike obstinacy: a forehead impudent, refusing to be ashamed; and then two eyes turned up seraphically languishing, as in divine contemplation and adoration; a touch of quiz too: on the whole, perhaps the most perfect quack-face produced by the eighteenth century." It was both natural and inevitable, Carlyle believed, that this man should have seized on the same opportunity for swindling as "Jeanne de Saint-Remy de Saint Shifty," the queen of liars, a sexual enchantress whose supreme quality was "untameableness." Whether as rivals or allies, Carlyle believed, Jeanne and Giuseppe had managed the diamond necklace affair to feed their own greed, and in the process they set in train a revolution that changed the world forever.

When in 1837 Carlyle published his greatest book, *The French Revolution*, he thought it natural to give the last scene to Cagliostro, the man who had prophesied the revolution and embodied its outcome: " '. . . all dwellings of men destroyed; the very mountains peeled and riven, the valleys black and dead: it is an empty World! Woe to them that shall be born then!— A King, a Queen . . . were hurled in . . . Iscariot Égalité was hurled in; thou grim de Launay, with thy grim Bastille; whole kindreds and peoples; five millions of mutually destroying Men. For it is the end of the dominion of IMPOSTURE (which is Darkness and opaque Firedamp); and the burning up, with unquenchable fire, of all the gigs that are in the Earth.' This prophecy, we say, has it not been fulfilled, is it not fulfilling?"

Carlyle's depiction of Cagliostro as an archetypal quack was so compelling that, like all myths, it soon broke free of its specific historical anchorage to float into our times. Here the comparably brilliant social critic, Umberto Eco, has caught and towed it to a new berth. For Eco, Cagliostro is prophetic, less of the eighteenth-century Enlightenment or the industrial age than of our own postmodern day. He sees Cagliostro as an empty sign, a person of such transparent ordinariness that he has become a magnet for the multiple fantasies of those who have lost any sense of reality. Today, he argues, adventurers like Cagliostro travel not so much to Italy or France as to California, where they don the gear of "new age" prophets, magicians, and healers and prey on people's psychological uncertainty and moral bewilderment. Today's Cagliostros declaim a "crisis of reason," substituting for the real world an "absolute fake industry"—hyperrealistic "castles of enchantment," "monasteries of salvation," kitsch cathedrals, and replica-filled cemeteries of reassurance. Today's modern shamans on television echo the Copt, raising their eyes upward to evoke divine visitation, grunting as they brawl with the devil and heal people through mellifluous words and stroking hands. In Los Angeles rather than Rome or Paris, we can find the direct lineal descendants of Egyptian Masonry, a hodgepodge of new age gurus, Internet clairvoyants, necromantic exorcists, pseudo Rosicrucians and latter-day Knights Templars.

The twentieth-century philosopher Walter Benjamin believed, however, that Cagliostro's role as a bearer of magic made him a titan in the history of western culture. Benjamin regarded Cagliostro as an underground messiah and a genius of disorder preaching creative irrationality in a repressive world. To Benjamin, who regarded men like Doria as forerunners of twentieth-century totalitarianism, Cagliostro was the healthy grit in the machine of the Enlightenment. Cagliostro stood for the repressed magical origins of science that had been forced underground after the French Revolution. He was the last true alchemist, a phantom of irrationality returning like the spirits at his séances to haunt the fetishists of reason.

Cagliostro has also become a bearer of modern-day dreams of magic. Lewis Spence's *Encylopedia of the Occult* calls him "one of the greatest occult figures of all time." Mediums claim to be his reincarnated spirit or sell his recipes for rejuvenation. The famous Swiss spiritualist and Mars-bound astral traveler Hélène Smith (Catherine Elise Muller) was said to acquire a double chin and a Sicilian accent when Cagliostro took charge of her soul. And the celebrated Russian mystic Madame Blavatsky (1831–1891) acknowledged Cagliostro as the true originator of her program of hermetic Egyptian theosophy. Nowadays the cinema has also replaced spiritualists like these as Cagliostro's reincarnater. Giuseppe Balsamo would have been delighted to see himself in the contemporary science fiction fantasy film *Spawn*, leading God's corps of magical gunslingers against necromantic monsters from the depths of hell.

Nothing demonstrates Cagliostro's ascension into culture better than the fact that his name has become a synonym for the word "magic" itself. In Webster's *Dictionary of Synonyms and Antonyms*, he shares this honor with Svengali. But the dictionary doesn't tell us whether the word "Cagliostro" stands for white or black magic, for the work of God or the work of the devil. Either way or both ways, Cagliostro has become an immortal. By evolving into an artistic figure, an aesthetic principle, and a popular icon,

Iain McCalman

Giuseppe Balsamo finally ascended, as he'd promised, to join the celestial spirits who rule over human destiny.

Of all the real and mythic lives of Giuseppe Balsamo—Alessandro Cagliostro, the one, finally, that attracts me most is the one I encountered first outside the house of Balsamo in Palermo—the one that launched my project of writing this book. It is the idea of Cagliostro as a flawed local hero. A contemporary journalist, Vincenzo Salerno, puts it neatly: "Every Italian city has its saintly protector.... Which historical figure could be said to be the embodiment of the Sicilian capital, the very archetypal exemplar of the 'Palermitanism' for which the city is famous today? The most obvious choice is the eighteenth-century charlatan, imposter, alchemist, archdeceiver, and all-around 'flim-flam' man, the so-called Count of Cagliostro." I think he's right. My own friends in Palermo—Sebastian, Nina, and the patron of the restaurant—were similarly realistic and similarly proud of Giuseppe Balsamo, the *uomo di popolo* from Palermo. Nina said it best—"Cagliostro may have been a crook, but he had a great soul."

PROLOGUE: THE HOUSE OF BALSAMO

3   *most profound veneration:* [Monsignor Barberi,] *The Life of Joseph Balsamo, commonly called Count Cagliostro: containing the Singular and Uncommon Adventures of that extraordinary personage, from his Birth till his Imprisonment in the Castle of Saint Angelo, translation of original proceedings published at Rome by Order of the Apostolic Chamber,* Dublin, 1792, pp. 68–69.

3   *fraud and superstition:* Thomas Carlyle, "Count Cagliostro: In Two Flights [1833]", *Critical and Miscellaneous Essays,* 5 vols., London, 1898, v. 3, pp. 298–301.

3   *paralyzed the will:* Baronne d'Oberkirch, *Mémoires sur la cour de Louis XVI,* 2 vols., Brussels, 1854, v. 1, pp. 2–9.

11 *the holy saints:* Barberi, *Life of Joseph Balsamo,* pp. 4–5.

13 *in every detail:* Alfred R. Weber, "Cagliostro in den Augen seiner Zeitgenossen," *Basler Jahrbuch,* Basel, 1959, p. 161.

13 *jewels of the hoard:* Roberto Gervaso, *Cagliostro: A Biography,* London, 1974, pp. 26–27.

I. FREEMASON

17 *Accepted Masons: A New English Dictionary on Historical Principles,* ed. J. A. H. Murray, Oxford, 1897, v. IV.

21 *badly needed entertainment:* Giacomo Casanova, *History of My Life*, 12 vols., ed. Willard R. Trask, London and Maryland, 1966, v. 11, pp. 156–167. For Casanova, see Lydia Flem, *Casanova or the Art of Happiness*, Harmondsworth, 1998; Derek Parker, *Casanova*, Stroud, 2002.

22 *fascination with alchemy:* Frederick Smith, *Brethren in Chivalry*, London, 1991, pp. 1–12; François Ribadeau Dumas, *Cagliostro*, Allen and Unwin, London, 1966, pp. 17–19; Gervaso, *Cagliostro*, pp. 28–29; W. R. H. Trowbridge, *Cagliostro: The Splendour and Misery of a Master of Magic*, New York, 1910, pp. 19–48; Jean Villiers, *Cagliostro: Le prophète de la révolution*, Paris, 1988, pp. 29–33.

23 *a secular habit:* Barberi, *Life of Joseph Balsamo*, pp. 12–13; Archives Nationales, Paris, Y/13125, MSS letter dated Palerme en Sicile, le 2, 9, 1786, Antoine Bracconiere to M. Fontaine, Commissaire.

24 *priestlike aura:* P. Maruzzi, *Bibliografia massonica italiana*, quoted in Gervaso, *Cagliostro*, p. 32.

24 *pilgrims visiting Rome:* Constantin Photiadès, *Les vies du Comte de Cagliostro*, Paris, 1932, pp. 93–95.

26 *letters of credit:* Archives Nationales, Y/13125. Pièces relative au procès du cardinal de Rohan, Interrogatoire of Laurence Feliciani, wife of Balsamo, Monsieur le Commissaire Bernard-Louis Philippe Fontaine, at Châtelet.

27 *beautiful blue eyes:* Casanova, *My Life*, v. 11, p. 166.

29 *eyes of God:* Casanova, *My Life*, v. 11, pp. 166–167.

29 *Antibes, to Aix:* Archives Nationales, Y/13125. Crime reports to Monsieur le Commissaire Bernard-Louis Philippe Fontaine, at Châtelet, Interrogatoires, Laurence Feliciani, 1, 11 February 1773.

30 *not a living:* Casanova, *My Life*, v. 11, p. 167.

30 *hard work:* Casanova, *My Life*, v. 11, p. 168.

31 *metropolis of London:* Photiadès, *Les vies de Cagliostro*, pp. 101–105.

33 *resumed their travels:* Archives Nationales, Y/13125. Interrogatoire, Laurence Feliciani, Bernard Louis Philippe Fontaine, Châtelet, 11 February 1773.

34 *never to return:* Gervaso, *Cagliostro,* pp. 43–44; Photiadès, *Les vies de Cagliostro,* pp. 122–123.

35 *elixir of life:* On the Wandering Jew, see George K. Anderson, *The Legend of the Wandering Jew,* California, 1965, esp. pp. 128–131; Brian Stableford, ed., *Tales of the Wandering Jew,* Cambridge, 1991, pp. 6–7.

35 *known Jesus personally:* Paul Chacornac, *Le comte de Saint-Germain,* Paris, 1989.

36 *to rescue him:* Marc Haven, *Le maître inconnu: Étude historique et critique sur la haute magie,* Lyon, 1964, pp. 33–48.

37 *Brother Hardivilliers: Courier de l'Europe,* 20 (33), 24 October 1786.

38 *upper-class males:* John Hamill, *The History of English Freemasonry,* Surrey, 1994, pp. 19–59.

38 *of their members:* Jasper Ridley, *The Freemasons,* London, 2000, pp. 47–67.

39 *sumptuous ceremonies:* Smyth, *Brethren in Chivalry,* pp. 13–22; Lewis Spence, *The Encyclopedia of the Occult,* London, 1988, pp. 405–408.

39 *Christian piety:* Albert G. Mackey, *Encyclopedia of Freemasonry and Its Kindred Sources,* Philadelphia, 1917, pp. 397–401, 482–491, 752–755, 865–866; Rosemary Ellen Gulley, *Harper's Encyclopedia of Mystical and Paranormal Experience,* New York, 1991, pp. 518–522.

40 *a considerable honor: Courier de l'Europe,* 20 (33), 24 October 1786.

41 *Martines de Pasqually:* Henry Ridgely Evans, *Cagliostro and His Egyptian Rite of Freemasonry,* Washington, 1919, pp. 12–17.

42 *Pellegrini and Cagliostro:* Barberi, *Life of Balsamo,* pp. 60–61.

42 *they were royals:* "Le Comte de Cagliostro. Exposition," 21 May–11 June 1989, Les Baux de Provence, Association culturelle Les Amis du Prince Noir, p. 78 (typescript, Bibliotheque Nationale, Paris).

43 *stones of Castille:* Giacomo Casanova, *Soliloque d'un penseur* [1786], *Bibliotecha Casanoviana,* v. 1, Paris, 1998, pp. 27–28.

44 *professional sycophant:* Giovanni Comisso, *Les Agents secrets de Venise, 1705–1797,* Paris, 1990, pp. 80, 154; S. Guy Endore, *Casanova: His Known and Unknown Life,* New York, 1929, pp. 271–293; Philippe Monnier, *Venise au XVIIIe siècle,* Paris, 2001, pp. 159–174.

45 *Maleficia—witchcraft:* G. Henningsen and John Tedeschi, eds., *The Inquisition in Early Modern Europe: Studies in Sources and Methods,* DeKalb, Ill., 1986, p. 135.

45 *polite or complaisant:* Casanova, *Soliloque d'un penseur,* pp. 29–30.

46 *Venus and Adonis:* Photiadès, *Les vies de Cagliostro,* p. 92.

46 *merchant went bankrupt:* Casanova, *Soliloque d'un penseur,* p. 37.

47 *a lodge in Paris:* Giacomo Casanova, *The Memoirs,* 6 vols., ed. Arthur Machen, New York, 1959–1960,v. 2, p. 102.

47 *public order:* Casanova, *The Memoirs,* v. 2, p. 99.

47 *against Freemasonry:* Casanova to Zaguri, 23 December 1792, "Supplement," *The Memoirs,* v. 6, p. 695.

47 *missions of recruitment:* Endore, *Casanova,* quoting Le Gras, pp. 178–179.

48 *bad acquaintances:* Casanova, *The Memoirs,* v. 2, p. 100.

2. NECROMANCER

49 *ghosts of the dead:* Samuel Johnson, *A Dictionary of the English Language,* London, 1756.

51 *Strict Observance Lodge:* Charlotte Elisabeth Constantia de la Recke [Elisa von der Recke], "Relation de séjour qu'a fait le fameux Cagliostro à Mitau en 1779 et des opérations magiques qu'il y a faites," "Introduction," Archiv der Familie Huber, Stat Archive, Basel, 694 A 6, pp. 3–4.

51 *disrupt Freemasonry:* Photiadès, *Les vies de Cagliostro,* pp. 98–101.

52 *in black magic:* Steven Luckert, "Cagliostro as a Conspirator: Changing Images of Joseph Balsamo in Late Eighteenth-Century Germany,"

Politica e Storia. Saggi e Testi, 43, Centro Editoriale, 1994, pp. 191–210.

54 *I have ever met:* Charlotte Elisabeth von der Recke, *Nachricht von des beruchtigten Cagliostro Augenthalte in Mitau, im Jahre 1779*, Berlin and Stettin, 1787, p. 26.

54 *imaginable on earth:* von der Recke, *Nachricht*, p. 26.

57 *Zobiachel, and Anachiel:* Photiades, *Les vies de Cagliostro*, pp. 165–166.

58 *the highest magic:* von der Recke, *Nachricht*, pp. 82–84; de la Recke, "Relation de séjour."

60 *can't read it:* von der Recke, *Nachricht*, p. 90.

60 *would be cleansed:* de la Recke, "Relation de séjour," p.61.

61 *to his superiors:* von der Recke, *Nachricht*, pp. 82–83, 86–87.

61 *part of the story:* von der Recke, *Nachricht*, pp. 5–6, 167.

61 *a passion for me:* von der Recke, *Nachricht*, p. 5.

62 *weeping inconsolably:* Johann Lorenz Blessig, *Leben des Grafen Friedrich von Medem nebst seinem Briefwechsel hauptsachlich mit der Frau Kammerherrin von der Recke, seiner Schwester*, Strasbourg, 1792, p. 54.

62 *inability to concentrate:* Blessig, *Friedrich von Medem*, pp. 9–13, 27–28, 51–52.

62 *weak and tearful:* Blessig, *Friedrich von Medem*, p. 32.

62 *he'd replied:* Blessig, *Friedrich von Medem*, pp. 105–107, 140.

63 *thoughts from me:* Blessig, *Friedrich von Medem*, pp. 105, 107.

63 *his "lust":* Friedrich Parthey to Elisa von der Recke, 18 May 1779, in *Elisa von der Recke, Tagebücher*, v. 2, Leipzig, 1902, p. 111.

63 *inner chagrin:* Blessig, *Friedrich von Medem*, pp. 138–139.

64 *to his death:* von der Recke, *Nachricht*, p. 46.

64 *shudders in me:* von der Recke, *Nachricht*, p. 54.

65 *its own sake:* von der Recke, *Nachricht*, pp. 60–62.

67 *cure the pain:* von der Recke, *Nachricht*, p. 72.

68 *this sublime science:* von der Recke, *Nachricht*, pp. 76–78.

68 *him in Russia:* von der Recke, *Nachricht*, p. 112.

69   *the same body*: von der Recke, *Nachricht*, pp. 137–143.

71   *ended in "el"*: von der Recke, *Nachricht*, pp. 130–134.

72   *well-being of mankind*: de la Recke, "Relation de séjour," p. 18.

3. SHAMAN

75   *other parts*: New English Dictionary, 1914, v. VIII, part 2.

77   *cabaret rope-dancer*: Charles Henri Heyking, "C. parmi les Russes," *Initiation*, 1898, p. 5.

78   *respond to my questions*: Heyking, "C. parmi les Russes," p. 8.

78   *he added quietly*: Heyking, "C. parmi les Russes," p. 9.

78   *the Spanish crown*: Initiation, supplement (August 1898), p. 133.

79   *Spanish nobleman*: Photiadès, *Les vies de Cagliostro*, p. 179.

79   *Foreign Affairs*: Casanova, *My Life*, v. 10, pp. 80–92.

80   *his bare hands*: Casanova, *My Life*, v. 10, p. 101.

80   *the same gardens*: Casanova, *My Life*, v. 10, pp. 99–150.

81   *avoided open politics*: For a sophisticated modern version of this argument, see Margaret C. Jacob, *Living the Enlightenment: Freemasonry and Politics in Eighteenth-Century Europe*, Oxford, 1991.

82   *clothed savages*: Henri Troyat, *Catherine the Great*, London, 2000, p. 63; Hedwig Fleischhacker, *Mit Feder und Zepter: Katharina II als Autorin*, Weinsberg, 1978; Douglas Smith, "Freemasonry and the Public in Eighteenth-Century Russia," *Eighteenth-Century Studies*, 29, (1), 1996, pp. 25–44.

83   *to see spirits*: Simon Sebag Montefiore, *Prince of Princes: The Life of Potemkin*, London, 2000, p. 209.

84   *of being discovered*: m. Barberi, *Life of Joseph Balsamo*, p. 72.

84   *with Russia's enemies*: Isabel de Madariaga, *Russia in the Age of Catherine the Great*, New Haven and London, 1981, p. 188.

84   *Masonic lodges*: L.-H.Labande (ed.), *Un diplomate français à la cour de Catherine II: Journal intime du Chevalier de Corberon, chargé d'affaires de France en Russie*, 2 vols., Paris, 1901, v. 1, p. 211.

84   *Prince Grigori Potemkin:* Montefiore, *Potemkin,* p. 349.

84   *knout or whip:* Montefiore, *Potemkin,* pp. 55–56.

85   *voyager of the spirits:* de Corberon, *Journal intimé,* v. 1, Introduction p. lxiv; Wilfred-Reneé Chettoui, *Cagliostro et Catherine II,* Paris, 1947, p. 35; Marc Haven, *Le maître inconnu Cagliostro: Étude historique et critique sur la haute magique,* Lyon, 1964, p. 71.

85   *give most attention:* von der Recke, "Relation du séjour," pp. 22–23.

86   *nervous disease:* de Corberon, *Journal intimé* p. 396.

86   *delicious distilled water:* De Corberon, *Journal intimé,* September 1780; Chettoui, *Cagliostro et Catherine,* pp. 42–43.

86   *positively beneficial:* Haven, *Le maître inconnu,* pp. 259–260.

87   *given up hope:* De Corberon, *Journal intimé,* p. 396.

88   *in wild fancies:* Clementio Vannetti, *L'évangile de Cagliostro: Liber memorialis de Caleostro cum esset Robereti,* reproduced in Haven, *Le maître inconnu,* pp. 288–289.

89   *simple human sentiment:* Chettoui, *Cagliostro et Catherine,* p. 37.

89   *to be the pig:* Vannetti, *Évangile de Cagliostro,* reproduced in Haven, *Le maître inconnu,* pp. 291–292.

91   *their lord:* Jean-Benjamin de La Borde, *Lettres sur la Suisse adressées à Madame de M*** par un voyageur Français, en 1781,* Genève, 1781; 17 June 1781, pp. 10–11.

92   *shaman for interrogation:* Chettoui, "Le chaman de Siberie," reproduced in *Cagliostro et Catherine,* pp. 181–182.

92   *the money and stay:* Montefiore, *Potemkin,* p. 209.

93   *you too much:* Montefiore, *Potemkin,* p. 174.

94   *honest conversation:* Catherine II, *Le secret de la société,* reproduced in Chettoui, *Cagliostro et Catherine,* pp. 197–198.

94   *specifically at him:* Haven, *Le maître inconnu,* pp. 76–77.

95   *What a king:* Richard Butterwick, *Poland's Last King and English Culture: Stanislaw August Poniatowski (1732–1798),* Oxford, 1998, pp. 78, 163–170, 219.

96 *Cagliostro's prediction:* de la Borde, *Lettres sur la Suisse,* letter iv, p. 14.

98 *the red poppy:* de la Borde, *Lettres sur la Suisse,* letter iv, p. 240.

99 *and a half ounces: Cagliostro démasqua à Varsovie, ou relation authentique de ses opérations alchimiques et magiques faites dans cette capitale en* 1780, *par un témoin oculaire,* Warsaw, 1786, p. 8.

100 *fabricating pearls: Cagliostro démasqua,* p. 24.

102 *made for kissing:* Montefiore, *Potemkin,* p. 29.

104 *secret society:* John Hamill and R. A. Gilbert, *World Freemasonry: An Illustrated History,* London, 1991, pp. 74–80; Una Birch, *Secret Societies of the French Revolution,* London, 1911, pp. 29–42.

104 *phantoms into flesh:* Gervaso, *Cagliostro,* pp. 92–93; Trowbridge, *Cagliostro,* p. 156; Barberi, *Life of Joseph Balsamo,* p. 75.

4. COPT

105 *Monophysites: New English Dictionary,* 1893, part 7.

108 *hear about it:* Frances Mossiker, *The Queen's Necklace,* London, 1961, p. 266.

108 *sort of a creature:* Mossiker, *The Queen's Necklace,* p. 300.

109 *Dutch china mug:* Jeanne de La Motte, *The Life of Jane de St-Rémy de la Motte,* 2 vols., Dublin, 1792, v. 1, pp. 106–107.

109 *his sovereign's wife:* Henry Vizetelly, *The Story of the Diamond Necklace,* London, 1887, pp. 275–276.

110 *ever really recover:* Madame Campan, *Memoirs of Marie-Antoinette,* 2 vols., London, 1909, v. 2, p. 23.

110 *other healing potions:* Gréte de la Francesco, *The Power of the Charlatan,* trans. Miriam Beard, New Haven, 1939, p. 214.

110 *from my eyes: Memorial for Count Cagliostro, Plaintiff, versus Maître Chesnon . . . Commissary in the Châtelet of Paris, and Le Sieur de Launay. . . Governor of the Bastille,* London, pp. 23-4.

111 *behind it all:* Abbé Georgel, *Mémoires pour servir à l'histoire des événements de la fin du dix-huitième siècle,* 2 vols., Paris, 1817, v. 2, p. 120.

111 *occult and supernatural:* Georgel, *Mémoires,* v. 2, p. 45.

111 *around 1,500 members:* Jacob, *Living the Enlightenment,* pp. 179–202.

111 *avoided him:* Photiadès, *Les vies de Cagliostro,* pp. 192–193.

112 *fellow-beings:* de La Borde, *Lettres sur la Suisse,* pp. 5–8.

113 *"House of the Virgin":* Haven, *Le Maître Inconnu,* pp. 87–93.

115 *paralyzed the will:* Mossiker, *The Queen's Necklace,* pp. 96–102.

115 *Tokay flow:* Vizetelly, *The Story of the Diamond Necklace,* pp. 76–77; Georgel, *Mémoires,* v. 2, pp. 50–51.

116 *feelings of my heart:* "Lettre de M. Sarasin, Négociant de Bâle, à M. Straub, Directeur de Manufacture Royale d'armes blanches, en Alsace. À Strasbourg,' 10 November 1781, Supplément au no. 365, *Journal de Paris,* 31 December 1781.

118 *penetrating wit: Mémoires du Comte Beugnot, ancien ministre publiés par le Comte Albert Beugnot, son petit-fils,* 2 vols., Paris, 1866, v. 1, p. 11.

119 *flare of passion:* Rétaux de Villette, *Mémoires,* quoted in Mossiker, *The Queen's Necklace,* p. 64.

119 *endeavored to attain:* de la Motte, *Life of Jane de St-Remy de la Motte,* v. 1, pp. 384–385.

119 *denied her birthright:* Beugnot, *Mémoires,* v. 1, p.11.

120 *with its anus:* Mossiker, *The Queen's Necklace,* pp. 114–115.

120 *complete discretion:* Vizetelly, *The Story of the Diamond Necklace,* pp. 37–39; Frantz Funck-Brentano, *The Diamond Necklace,* trans. H. S. Edwards, London, 1911, pp. 76–79; Mossiker, *The Queen's Necklace,* pp. 113–115.

121 *should not be done:* Beugnot, *Mémoires,* v. 1, p. 27.

124 *news to his wife:* "Interrogatoire de la fille Leguay, dite D'Oliva," Émile Camperdon, *Marie-Antoinette et le procès du collier, d'après la procedure instruite devant le Parlement de Paris,* Paris, 1863, pp. 351–356.

124 *fell to her neck:* Beugnot, *Mémoires,* v. 2, p. 89.

125 *your grassy seat:* "Justificatory Pieces," No. XIII, Letter from the Cardinal to the Queen, 29 July 1784, p. 23, appended to *Memoirs of the Countess de Valois de La Motte,* London, 1789.

126 *trap was sprung:* Georgel, *Mémoires*, v. 2, p. 60.

127 *his only reward:* Gervaso, *Cagliostro*, p. 127.

128 *covered the walls:* G. Lenôtre, "La Maison de Cagliostro," *Vielles maisons, vieux papiers*, Paris, 1912, pp. 161–171.

128 *religion and magic:* Evelyn Farr, *Before the Deluge; Parisian Society in the Reign of Louis XVI*, London and Chester Springs, 1994, pp. 117–122, 126–127.

129 *sexual titillation:* Funck-Brentano, *The Diamond Necklace*, pp. 99–100.

129 *gossip and drinking:* Barberi, *Life of Joseph Balsamo*, pp. 136–143; Gervaso, *Cagliostro*, pp. 128–132, quoting from the *Leyden Gazette*.

129 *delicate an operation:* Archives Nationales: Fonds Parlement de Paris, x/2B/1417, Trial of Cardinal Rohan, 30 January 1786, Interrogatoire de Comte Cagliostro by Jean-Baptiste Maximilian Pierre Titon.

129 *fitted the bill perfectly:* Beugnot, *Mémoires*, v. 1, p. 64.

130 *27 March 1785:* Dossier, "Affaire du Collier: Pièces concernant les S et D, Cagliostro, historique et divers particuliers impliques dans cette affaire," Interrogatoire, 24 August 1785, Interrogatoire du Sieur de Cagliostro, 30 January 1786, Camperdon, *Marie-Antoinette*, pp. 342–346.

130 *into the abyss:* Beugnot, *Mémoires*, v. 1, pp. 61–62.

131 *new mistress:* Funck-Brentano, *The Diamond Necklace*, p. 304.

131 *collective orgy:* 'Echte Nachtrichten von dem Grafen Cagliostro. Aus der Handschrift seines enhflohenen Kammerdieners', from Klaus H. Keifer, *Cagliostro: Dokumente zu Auflarung und Okkulismus*, Frankfurt 1991, pp. 245-80.

132 *across the room: Mémoire pour le Cardinal de Rohan*, pp. 48–49.

133 *bed to sleep on: Life of Jane de St. Rémy*, v. 1, pp 340–363.

133 *sell in Britain:* Premier interrogatoire de Madame de la Motte…21 January 1786, Camperdon, *Marie-Antoinette*, pp. 312–317.

134 *earlier magicians: Histoire de collier ou Mémoire de la comtesse de la Motte contre*

M. Le Cardinal de Rohan et du soi-disant Comte de Cagliostro, Paris, 1786, pp. 27–28.

134 *this latest mistress:* Réponse pour la comtesse de Valois La Motte, au mémoire du comte de Cagliostro, Paris, 1786, pp. 3–36.

134 *buy the memoir:* Life of Jane de St-Rémy, v. 1, pp. 375–377.

135 *much the better:* Life of Jane de St-Rémy, v. 1, pp. 359–361.

135 *dies adoring you:* Reproduced in Photiadès, Les vies de Cagliostro, pp. 298–299.

136 *for healing services:* Interrogatoire du Sieur de Cagliostro 25 August 1792, Dossier, Affaire du Collier, AN F7/4445 and 4550/2; Interrogatoire du Sieur de Cagliostro, 30 January 1786, in Camperdon, Marie-Antoinette, pp. 338–350.

136 *that will speak:* Memoirs of the Countess de Valois de la Motte, London, 1789, pp. 238–239.

137 *history of my life:* Memorial, or Brief, for the Count Cagliostro, Defendant, against the Attorney-General, Plaintiff, in the cause of Cardinal Rohan, Comtesse de La Motte and others, London, 1786, p, 2.

138 *queens and whores:* Funck-Brentano, The Diamond Necklace, pp. 283–285; Robert Darnton, The Literary Underground of the Old Regime, Cambridge, Mass. and London, 1982, chs. 1, 6; Robert Darnton, The Forbidden Bestsellers of Pre-Revolutionary France, London, 1996; Sarah Maza, Private Lives and Public Affairs: The Causes Célèbres of Prerevolutionary France, Berkeley, Los Angeles and London, 1993, pp. 167–211.

139 *care or consolation:* "Requête au Parlement, les chambres assembleés, par le Comte de Cagliostro...24 Février 1786," Bibliothèque de l'Arsenal, Box MSS 12457.

140 *rigors of the Bastille:* Gréte de Francesco, The Power of the Charlatan, pp. 214–215.

141 *them no more:* Memorial for Count Cagliostro, pp. 29–32

142 *sin of the Salpêtrière:* Jeanne de La Motte, Life of Jane de St-Rémy, v. 1, pp. 125–127.

5. PROPHET

143 *or generally:* New English Dictionary, 1909, v. VII, part 2.

143 *his fellow creatures:* Universal Register (The Times), 7, 30 July 1786.

143 *the lowest mechanic:* Universal Register (The Times), 15 March 1786.

144 *the Turkish Empire:* Parkyns Macmahon, "Introductory Preface," *Memorial, or Brief for the Comte de Cagliostro,* London, 1786, p. xi. On Macmahon and his journalist wife, see Pierre Manuel, *La police de Paris dévoilée,* 2 vols., Paris, 1791, v. 1, pp. 248–249; *Correspondence of Hugh Walpole,* ed. W. S. Lewis, Oxford, 1971, v. 25, pp. 557, 631; British Library, Egerton MS, 3438, Parkyns Macmahon to Lord Holdernesse, 29 May 1758, ff. 245–246, 5 October 1758, ff. 328–329; *Times* (obituary), 16 January 1788.

144 *hoped to influence:* Times, 7 July 1786. See also 7, 15, 20 March; 24 April; 29, 30 June; 6, 7, 8 July. Simon Burrows, *French Exile Journalism and European Politics, 1792–1814,* Suffolk and Rochester, 2000, pp. 8, 74, 165.

145 *the Egyptian Rite:* Universal Register (Times), 8 May 1786; Marsha Keith Schuchard, "Lord George Gordon and Cabalistic Freemasonry. Beating Jacobite Swords into Jacobin Ploughshares," author's manuscript, p. 26, forthcoming in *Secret Conversions to Judaism in Early Modern Europe,* eds. Martin Mulsow and Richard Popkin.

145 *Nobility of Paris:* Gordon P. G. Hills, "Notes on the Rainsford Papers in the British Museum," *Ars Quatuor Coronatorum* [AQC], xxxvii (1913), pp. 93–117; Gordon P. Hills, "Notes on Some Masonic Personalities at the End of the Eighteenth Century," AQC, xxv (1912), pp. 141–164; Marsha Keith Schuchard, "The Secret History of Blake's Swedenborgian Society," *Blake / An Illustrated Quarterly* (fall 1992), pp. 40–51.

146 *Stockholm and Berlin:* Joscelyn Godwin, *The Theosophical Enlightenment,* New York, 1994, pp. 93–114.

146 *alchemical transmutation:* Other than the writings of Schuchard above,

see also Clarke Garrett, *Respectable Folly*, Baltimore and London, 1975, especially ch. 5; Peter Ackroyd, *Blake*, London, et al, 1995, especially chs. 15–16.

146 *gardens in Knightsbridge*: On Swinton, see the anonymous biography evidently written by a disciple of Cagliostro, *The Life of Count Cagliostro*, London, 1787, pp. 94–95; Margery Weiner, *The French Exiles, 1789–1815*, London, 1960, pp. 119–120.

147 *own needs*: Sophie von La Roche, *Sophie in London, 1786: Being a diary...*, ed. Clare Williams, London, 1933, pp. 136–141, 148–150.

147 *body and soul*: *Lettre du Comte de Cagliostro au peuple Anglais: Pour server de suite à ses mémoires*, London, 1786, pp. 41-2.

148 *for these villains*: Paul Robiquet, *Théveneau de Morande: Étude sur le XVIII siècle*, Paris, 1882, pp. 9–21.

149 *those they libeled*: Robiquet, *Théveneau de Morande*, pp. 62–65; Trowbridge, *Cagliostro*, pp. 260–262; Darnton, *Literary Underground*, pp. 1–40.

149 *hungry children*: Gary Kates, *Monsieur d'Eon Is a Woman: A Tale of Political Intrigue and Sexual Masquerade*, New York, 1995, p. 213.

149 *Le philosophe cynique* (1771): Peter Wagner, *Eros Revived: Erotica of the Enlightenment in England and America*, London, 1990, pp. 92–99.

149 *writhed among the pages*: M. Dorothy George, *Catalogue of Personal and Political Satires. Preserved in the Department of Prints and Drawings in the British Museum*. 11 vols., London, 1935, v. 5, no. 5247, "The French Lawyer in London," p. 178.

149 *nothing but laugh*: Claude Manceron, *Toward the Brink, 1785–1787. The Age of Revolution*, 5 vols., New York and London, 1989, v. 4, p. 229.

150 *Chevalier d'Eon*: Kates, *Monsieur d'Eon*, pp. 210–214, 243–245.

150 *the French crown*: Kates, *Monsieur d'Eon*, pp. 213–214.

150 *the old regime*: Robert Darnton, *The Corpus of Clandestine Literature in France, 1769–1789*, New York and London, 1995, p. 199.

150 *good deal less*: Gervaso, *Cagliostro*, pp. 167–168.

151 *cross-channel smut trade:* Simon Burrows, "A Literary Low-Life Reassessed: Charles Théveneau de Morande in London, 1769–1791," *Studies in the Eighteenth Century,* 10, *Eighteenth Century Life,* 22, n.s. 1 (February 1998), pp. 83–85.

151 *at the chance:* Maza, *Private Lives and Public Affairs,* pp. 191–211.

152 *good political polishing:* Bibliothèque de l'Arsenal, Box MSS 12,457, "Copie d'une lettre écrite par Cagliostro a M.... 20 juin 1786," f. 21 ff.

153 *he obliged them:* Hans Jurgen Lusebrink and Rolf Reichardt, *The Bastille: A History of a Symbol of Despotism and Freedom,* Durham and London, 1997, pp. 8–37.

153 *that Miltonic hell:* Bibliothèque de l'Arsenal, Box MSS 12,457, "Copie d'une lettre écrite par Cagliostro a M...." The letter is reprinted in full in F. Ribadeau Dumas, *Cagliostro,* Paris, 1966, pp. 176–178.

155 *written to his sister:* *The Letters of Edward Gibbon: Volume Two, 1774–1782,* ed. J. E. Norton, London, 1956, pp. 243–245.

155 *modern man of terror:* See Iain McCalman, "Lord George Gordon and Madame La Motte: Riot and Sexuality in the Genesis of Burke's Reflections on the Revolution in France," *Journal of British Studies,* 35 (July 1996), pp. 343–367.

155 *with a penknife:* See, for example, *Times,* 7 September, 8 November, 11 November 1785; George, *Personal and Political Satires,* vol. 6, no. 7134, 30 January 1787, p. 393; no. 8249, 1 October 1785; Public Record Office, PC 1/3127, "A Letter of His Grace Archbishop of Canterbury," London, 1787, pp. 5–7.

156 *Ambassador d'Adhémar:* Munro Price, *Preserving the Monarchy,* Cambridge, 1995, pp. 182–183; Vizetelly, *The Story of the Diamond Necklace,* pp. 284–287; La Motte, *Memoirs,* pp. 158–178 (account inserted by Count La Motte).

157 *a tyrannical government:* "Information against Lord George Gordon. Defamatory Libels against Princess Marie-Antoinette and François Barthelemy," Michaelmas 27, George III, 1786, Public Record Office,

Treasury Solicitor's Files, TS 11/388/1212. See also *Public Advertizer* and *Morning Chronicle*, 22, 24, 26 August 1786.

157 *the house in Sloane Street*: See *Times*, 1 November 1786, for example, when Lord Bute was denied an interview on the grounds that he was not a friend of Gordon.

157 *the Great Copt*: Sarasin Archiv, 212 F 11, 27, Letters from Sophie von La Roche to Gertrude Sarasin, 30 October 1784; 2 August, 17 November 1785; 5, 23, March, 17 July 1786.

158 *his Lordship dislikes*: J. Paul de Castro, *The Gordon Riots*, London, 1926, p. 247.

158 *futile rubbish daily*: von La Roche, *Sophie in London*, pp. 136–149, 148–150.

158 *a substantial payment*: Haven, *Le maître inconnu*, pp. 186–187.

158 *any means possible*: Photiadès, *Les vies de Cagliostro*, p. 333.

158 *the French embassy*: Barberi, *The Life of Joseph Balsamo*, London, 1787, pp. 95, 102–104; *Lettre du Comte de Cagliostro au peuple Anglais, pour servir de suite à ses mémoires*, London, 1786, pp. 40–42.

159 *were promised*: *Courier de l'Europe*, 22 August 1786, 20 (15), p. 117; 29 August 1786, 20 (17), pp. 125–134; 1 September 1786, 20 (18), pp. 142–143.

159 *reared on arsenic*: *Courier*, 22 August 1786.

160 *pleased to try it*: *Courier*, 5 September 1786, 20 (19), p. 151.

160 *wish to dispatch*: *Lettre . . . au peuple Anglais*, pp. 53–55.

160 *Carlo Saachi*: *Courier*, 29 August and 1, 5 September 1786.

160 *literary glory*: von der Recke, *Tagebücher*, v. 2, July 1779, pp. 112–116; von der Recke, *Nachricht*, p. 154; von der Recke, *Mein Journal . . . 1791 und 1793–1795*, Leipzig, 1927, S. 77f.

161 *the same person*: *Courier*, 10, 13, 17, October 1786.

162 *and sound technology*: Richard Altick, *The Shows of London*, Camb., Mass, and London, pp. 119–127; Rudiger Joppien, *Philippe Jacques de Loutherbourg, RA, 1740–1812*, Kenwood, 1973; Morton D. Paley, *The*

Apocalyptic Sublime, Newhaven and London, 1986, pp. 51–70; Christopher Baugh, *Theatre in Focus: Garrick and de Loutherbourg*, Cambridge, and Alexandria, Va., 1990, pp. 21–51.

162 *eerie sound effects:* [W. H. Pyne,] *Wine and Walnuts; or After Dinner Chit-Chat by Ephraim Hardcastle*, 2 vols, London, 1823, v. 1, pp. 302–303.

162 *"phantasmagoria":* Terry Castle, "Phantasmagoria: Spectral Technology and the Metaphorics of Modern Reverie," *Critical Inquiry* (autumn 1988), pp. 26–61; Barbara Stafford, *Artful Science: Enlightenment Entertainment and the Eclipse of the Visual Education*, Cambridge, Mass., and London, 1994, pp. 74–79.

162 *the modern cinema:* Guy Chapman, *Beckford*, New York, 1937, pp. 99, 105–106.

162 *his medicines:* Paley, *Apocalyptic Sublime*, p. 52.

163 *Lodge in 1783:* Godwin, *Theosophical Enlightenment*, p. 100.

163 *occult libraries:* Jacques Philippe de Loutherbourg, *A Catalogue of All the Valuable Drawings, Sketches, Sea Views and Studies of That Celebrated Artist Jacques Philippe de Loutherbourg RA . . . Also Includes Library of Scarce Books*, London, 18 June 1812.

163 *Elizabeth Howard:* Photiadès, *Les vies de Cagliostro*, pp. 346–347.

163 *imponderable fluid: The Lectures of J. B. de Mainauduc, M.D., Member of the Royal College of Surgeons in London*, London, 1788; Patricia Fara, "An Attractive Therapy: Animal Magnetism in Eighteenth-Century England," *History of Science*, 32 (1995), pp. 138–145; Marsha Keith Schuchard, "Blake's Healing Trio: Magnetism, Medicine and the Mania," *Blake: An Illustrated Quarterly*, 22 (1989), pp. 20–22; G. E. Bentley Jr., "Mainaduc (sic), "Magic and Madness: George Cumberland and the Blake Connection," *Notes and Queries*, 236 (Sept.1991), pp. 294–296.

163 *England and abroad:* Marsha Keith Schuchard, "William Blake and the Promiscuous Baboons: A Cagliostrean Séance Gone Awry," *British Journal for Eighteenth-Century Studies* (autumn 1995), pp. 51–71.

164 *made for himself*: Cagliostro démasqua à Varsovie, Warsaw, 1786, p. 52.

165 *tinted clouds*: Courier, 19 October 1786, pp. 255–256, 26 October 1786, and 26 November 1786, pp. 270–272.

168 *Masonic rivals*: I am grateful to Dr. Michael Rhodes, Principal Curator of Torquay Council Gallery, for allowing me to make photographic reproductions of these wonderful, little-known paintings.

168 *endure all evening*: Photiadès, Les vies de Cagliostro, p. 343.

169 *a fairground quack*: Courier, 3 November 1786, pp. 290–291.

172 *put an end*: "Anecdote Maçonnique: A Masonic Anecdote," 11 November 1786, in George, *Personal and Political Satires*, pp. 332–334; General Advertizer, 29 November 1786.

172 *in the past*: Lettre du Cagliostro à peuple Anglais, especially pp. 1–38, 57–66.

172 *disappeared into hiding*: "Lord George Gordon and Madame La Motte," Journal of British Studies, 35 (July 1996), pp. 343–367.

173 *an ornate silver cane*: Courier, 13 October 1786, pp.237–238.

6. REJUVENATOR

175 *fresh again*: New English Dictionary, 1910, v. VIII, part 1.

176 *more than 200,000 guilders*: This was equivalent in the 1930s to more than 2 million Swiss francs. Lionel Gossman, *Basel in the Age of Burckhardt: A Study in Unseasonable Ideas*, Chicago and London, 2000, p. 45.

178 *repay them abundantly*: Raymond Silva, *Joseph Balsamo, alias Cagliostro*, Geneva: Ariston, 1975, p. 177.

179 *soaking up nature*: Sarasin Familien Archiv (Basel, Staat Archiv), 212 F 11, 23, Letters from Jacques Sarasin, 10–12 June 1783.

179 *many others*: Haven, *Le maître inconnu*, pp. 202–203; Antoine Faivre, *Kircherberger et l'illuminisme du dix-huitième siècle*, Paris, 1966, pp. xx–xxi, 78; Gossman, *Basel in the Age of Burckhardt*, pp. 204–205.

180 *reborn as immortals*: Barberi, *Life of Joseph Balsamo*, pp. 143–146.

181 *exhaustion of childbirth:* J. and G. Sarasin, "Rezeptbuchlein" (MS prescription book), Sarasin Archiv, 212, F 5, 49 pp.

181 *all over Europe:* S. Wildermett to J. Sarasin, 26 January 1787, Sarasin Archiv, 212 FII, 14, no. 5, and A. Wildermett to J. Sarasin, 10 March 1787.

182 *Cagliostro's recipes:* Sarasin Archiv, 212 FII, 14, letters from Sigismund Wildermett to J. Sarasin, 25, 26 January; 6, 7, 23 February; 8, 23, 31 March; 2 April; 4, 12, 29 May 1787.

182 *an exciting playmate:* Haven, *Le maître inconnu*, pp. 202–205.

182 *tumult of the world:* Sarasin Archiv, 212, FII, 33, Gertrude Sarasin, 23 June 1786, f. 14.

184 *states against Cagliostro:* Sarasin Archiv, 212, FII, 14, S. Wildermett to J. Sarasin, 23 July 1787.

185 *the universal panacea:* Archives russes, n.d. Quoted in Chettoui, *Cagliostro et Catherine*, p. 49.

185 *local journals:* Chettoui, *Cagliostro et Catherine*, p. 55.

186 *allows it:* Catherine II, "Character of the Shaman," in Chettoui, *Cagliostro et Catherine*, p. 57.

186 *claims for damages:* Sarasin Archiv, 212 FII, 4, 12, P. J. de Loutherbourg to S. Wildermett, n.d.

188 *the petticoat:* Sarasin Archiv, 212 FII, 33, 20, J. Sarasin to de Gingin, 12 January 1788.

188 *the two households:* Sarasin Archiv, 212 FII, 14, S. Wildermett to J. Sarasin, 9 October 1787.

188 *attacked by bile:* Sarasin Archiv, 212, FII, 4, 12, P. J. de Loutherbourg to J. Sarasin, n.d. (c. 3 October), 12, 14 October 1787.

189 *all her heart:* Sarasin Archiv, 212, FII, 33, 5, Suzette Sarasin to her parents, 7 January 1788.

189 *a plausible manner:* Sarasin Archiv, 3, 20, J. Sarasin to Gingin (brother-in-law), January 1788.

189 *to ruin Cagliostro:* Sarasin Archiv, 212, F11, 33, 5, Suzette Sarasin to her parents, 7 January 1788.

189 *to a duel:* Sarasin Archiv, 212 F11, Daschelhofer to J. Sarasin, 15 October 1787.

189 *claws of his enemies:* Sarasin Archiv, 212, F 11, 14, S. Wildermett to J. Sarasin, 6, 12 January 1788.

189 *served his master well:* Sarasin archiv, 212 F11, 33, "Report of Jacques Sarasin on the negotiations at Bienne and the arrangement between Count Cagliostro and P. J. de Loutherbourg," 12 January 1788.

190 *poniard to me:* Barberi, *Life of Joseph Balsamo,* pp. 196–197.

190 *had often predicted:* Sarasin Archiv, 212 F 11, 14, S. Wildermett to Gertrude Sarasin, 20 July 1789. See also Photiadès, *Les vies de Comte Cagliostro,* pp. 357–358.

190 *point of death:* Sarasin Archiv, 212 F11, 14, S. Wildermett to J. Sarasin, 21 January 1788.

190 *suffering from hysteria:* Sarasin Archiv, 212, F 11. 14, S. Wildermett to J. Sarasin, 15 February 1788.

191 *at their side:* Sarasin Archiv, 212, F 11, 14, S. Wildermett to J. Sarasin, 28 January, 3 February 1788.

192 *different from our own:* Sarasin Archiv, 212 F11, 14, S. Wildermett to J. Sarasin, 24 June 1788.

193 *by his unhappiness:* Sarasin Archiv, 212, F11, 14, S. Wildermett to J. Sarasin, 24 July 1788.

194 *before the house:* Clementino Vannetti, "L'évangile de Cagliostro," reprinted in Haven, *Le maître inconnu,* V, pp. 277.

194 *red of a ruby:* Vannetti, "L'évangile de Cagliostro,", p. 278.

194 *a different life:* Vannetti, "L'évangile de Cagliostro," X, p. 283.

194 *he will be punished:* Photiadès, *Les vies de Cagliostro,* pp. 367–368.

195 *the scriptures attentively:* Vannetti, "L'évangile de Cagliostro," XIV, p. 290.

195 *from her mouth:* Vannetti, "L'évangile de Cagliostro," XIV, p. 290; XV, p. 293.

196 *himself for his flock:* Vannetti, "L'évangile de Cagliostro,", XIV, p. 289.

197 *peace and security:* Photiadès, *Les vies de Cagliostro*, pp. 369–370; Vannetti, "L'évangile de Cagliostro," XI, p. 284.

198 *Cardinal Rohan:* A. Gagnière, "Cagliostro et les franc-maçons devant L'Inquisition," *Nouvelle Revue*, 43 (1886), pp. 40–42; Gervaso, *Cagliostro*, p. 202.

198 *as a secretary:* Photiadè's, *Les vies de Cagliostro*, pp. 381–382.

199 *a street fighter:* Photiadès, *Les vies de Cagliostro*, p. 384.

200 *you must adore:* Gervaso, *Cagliostro*, pp. 210–212.

200 *carefully noted:* Gervaso, *Cagliostro*, pp. 210–212; D. Dalbien, *Le Comte de Cagliostro*, Paris, 1983, p. 256.

201 *outside his room:* Dalbian, *Le Comte de Cagliostro*, p. 255–258.

202 *arbitrary royal edict:* Photiadès, *Les vies de Comte Cagliostro*, pp. 387–388; Raymond Silva, *Joseph Balsamo alias Cagliostro*, Geneva, 1975, p. 194.

202 *excellent manual skills:* Gervaso, *Cagliostro*, pp. 223–224; Photiadès, *Les vies de Comte Cagliostro*, p. 389.

202 *with the results:* Photiadès, *Les vies de Cagliostro*, pp. 389–390.

203 *by the guards:* Photiadès, *Les vies de Cagliostro*, pp. 397–398.

204 *devotion touched them:* Barberi, *Life of Joseph Balsamo*, pp. 217–218.

204 *even upon sinners:* Barberi, *Life of Joseph Balsamo*, pp. 205–207.

7. HERETIC

207 *catholick church:* Johnson, *Dictionary.*

208 *outside world:* State Archives of Pesaro, "A Collection of Letters pertaining to Giuseppe Balsamo known as Count Cagliostro, sentenced to imprisonment in the San Leo fortress by decree of His Holiness, Pope Pius VI," Book I, F. D. Cardinal de Zelada, Rome, to G. Cardinal Doria, Legate of Urbino, 16 April 1791, f. 1; Secretariat

of State, Rome, to Castellano, Sempronio Semproni, San Leo, 16 April 1791, ff. 1–2.

208 *Monsignor Giovanni Barberi*: Barberi, *Life of Balsamo*.

208 *legal expert, Roverelli*: Barberi, *Life of Balsamo*, especially pp. 178–269; Le Comte de Cagliostro. Exposition 21 May–11 June 1789, Les Baux de Provence, Association Culturelle Les Amis du Prince Noir, pp. 78–82.

211 *for that possibility*: Photiadès, *Les vies de Cagliostro*, pp.395–396.

211 *the papal states*: Photiadès, *Les vies de Cagliostro*, pp. 405–406.

211 *April 1791*: Gervaso, *Cagliostro*, pp. 189–228.

211 *key papal towns*: Gervaso, *Cagliostro*, p. 209.

212 *for their arrival*: G. Cardinal Doria to de Zelada, 22 April 1791, ff. 7–8.

214 *afraid he'd die*: Monsignor Terzi, Bishop of Montefeltro [Rimini], to Doria, 26 April 1791, ff. 23–24.

214 *kinds of iniquities*: G. Cardinal Doria, Pesaro, to Monsignor Terzi, Bishop of Montefeltro, 30 April 1791, ff. 23–24.

215 *and the prison*: G. Cardinal Doria to the Commissary of Montefeltro [23] April 1791, ff. 10–11.

216 *live without toil*: Tenente Gandini to Cardinal Doria, 22 April 1791, ff. 13–17.

216 *of the sovereign*: G. Cardinal Doria to Cardinal de Zelada, 24 April 1791, ff. 20–22.

217 *his family table*: G. Cardinal Doria to Cardinal de Zelada, 9 June 1791, f. 47.

217 *enormous consequences*: G. Cardinal Doria to Castellan Semproni, 4 May 1791, ff. 26–29.

217 *to be scrutinized*: G. Cardinal Doria to Castellan Semproni, 22 May 1791, ff. 38-9.

218 *to the Vatican*: G. Cardinal Doria to Castellan Semproni, 4 May 1791, f. 30.

218 *reformation is true*: G. Cardinal Doria to Cardinal de Zelada, 5 June 1791, ff. 47–48.

218 *taken care of*: G. Cardinal Doria to Cardinal de Zelada, 12 May 1791, ff. 35–36.

218 *by what means*: [Cardinal Doria] to Cristoforo Beni, n.d., ff. 55–56.

219 *into the fortress*: Cardinal Doria to Cardinal de Zelada, 3 July 1791, ff. 56–57.

219 *City of San Leo*: Cardinal Doria to Cardinal de Zelada, 10 July 1791, ff. 58–61.

219 *seen by anyone*: Doria to Semproni, n.d., f. 133.

219 *with the prisoners*: Pesaro State Archives, Book 2, Semproni to Doria, 30 July 1791, no. 41.

220 *count's present sufferings*: Book 2, Gianfrancesco Arrigone, Macerata, to Cardinal Doria, 5 August 1791, no. 42.

220 *in the Tesoro*: Book 2, Semproni to Cardinal Doria, 9 August 1791, no. 43.

220 *form of transport*: Doria to de Zelada, n.d., f. 138.

220 *design and navigation*: Clementino Vannetti, *The Gospel According to Count Cagliostro*, reproduced in Haven, *Le maître inconnu*, p. 273.

221 *published in our time*: Book 2, Semproni to Cardinal Doria, 16 August 1791, no. 45.

221 *surprise or deceived*: Cardinal Doria to Cardinal de Zelada, 28 September 1791, ff. 75–77.

222 *inefficiency and corruption*: Cardinal Doria to Cardinal de Zelada, 5 June 1791, f. 47.

222 *peace and quiet*: Cardinal Doria to Cardinal de Zelada, 27 September 1791, ff. 79–80.

222 *to this task*: Cardinal Doria to Castellan Semproni, 15 October 1791, f. 85.

223 *could be lethal*: Cardinal Doria to Cardinal de Zelada, 30 October 1791, f. 90.

223 *plotting to escape*: Cardinal Doria to Cardinal de Zelada, 10 November 1791, f. 94.

223 *make him submissive:* Cardinal Doria to Cardinal de Zelada, 27
November 1791, ff. 100–101.

224 *this detestable prisoner:* Doria to Gandini, 13 March 1792, ff. 118–119;
Doria to de Zelada, 1, 4 March 1792, ff. 119–122.

224 *his cell frequently:* Doria to Semproni, 21 May 1791, ff. 37–38.

224 *surprise and deceive:* Doria to Monsignor Terzi, 28 May 1791, ff. 42–43.

225 *them off guard:* Doria to Semproni, 27 August 1791, ff. 72–73.

225 *to be careful:* Doria to Semproni, 11 October 1791, ff. 84-5; 15 October
1791, ff. 85–86.

225 *himself to death:* Doria to Zelada, 16 October 1791, f. 86; Doria to
Semproni, 23 October 1791, ff. 87–88.

225 *of cell searches:* Doria to de Zelada, 30 October 1791, ff. 90–91;
10 November 1791, ff. 92–94.

226 *procure his reformation:* Doria to Semproni, 19 November 1791,
ff. 96–97.

226 *loathsome drawings:* Doria to de Zelada, 27 November 1791,
ff. 100–101.

226 *a true Catholic:* Doria to Zelada, 27 November 1791, ff. 100–101.

226 *an obstinate misbeliever:* Doria to de Zelada, 23 February 1792, f. 117.

226 *new confessors:* Doria to de Zelada, [August 1792,] ff. 139–140.

227 *or a satirical mind:* Doria to de Zelada, 27 November 1791.

227 *prophesies to Rome:* Doria to de Zelada, 31 March 1793, f. 23.

228 *Balsamo's deceitfulness:* Doria to de Zelada, 29 January 1792,
ff. 113–114.

228 *significant transgression:* Doria to Semproni, 2 July 1793, f. 28.

228 *the sovereign orders:* Doria to Semproni, 13 July 1793, f. 30.

228 *with this affair:* Doria to de Zelada, 26 January 1794, ff. 34-6.

230 *gate keys:* Doria to Semproni, 16 June 1792, f. 133.

230 *artillery, and ammunition:* Doria to de Zelada, 13 December 1792, f. 15.

230 *seemed to care:* Doria to de Zelada, 3 February 1793, f. 18.

230 *Corporal Marini:* Doria to de Zelada, 14 March 1793, ff. 21-3.

230 *dangerously exposed:* Doria to Semproni, 22 June 1793, f. 26.

231 *the government:* Doria to Semproni, 3 November 1792, ff. 11–12.

231 *insolence than usual:* Uditori to Semproni, 27 May 1794, f. 38.

231 *scandalous frenzy:* Uditori to de Zelada, 10 July 1794, f. 38.

231 *impenitent man:* Ferdinando Archbishop of Catagine, n.d., f. 43.

232 *shrewdness and ability:* Gervaso, *Cagliostro*, pp. 239–242.

EPILOGUE: IMMORTAL

233 *perpetual:* Johnson, *Dictionary.*

233 *convert his enemies:* Sarasin Archiv, Basel, 212, F11, 31, J. Sarasin to
Comte d'Estillac, February 1791, 12 April 1791, also F11, 32, J.
Sarasin to Cardinal Rohan, 5 February 1791.

234 *Report of 1786:* Klaus H. Kiefer, "Fiction et réalité: Aspects de la
réception de Cagliostro du 18e siècle à nos jours," *Presenza di Cagliostro:
Testi riuniti a cura di Danilea Gallingani,* Florence, 1994, pp. 423–453.

234 *Frenchmen would free him:* Elisa von der Recke, *Mein Journal:* 1791,
1793–1795, ed. Johannes Warner, Leipzig, 1927, pp. 161–162.

235 *Freemasonry in Russia:* Isabel de Madariaga, *Russia in the Age of Catherine
the Great,* Newhaven and London, 1981, pp. 522–530.

235 *several countries:* von der Recke, *Mein Journal,* pp. 276–277.

235 *French Revolution:* Augustin Barruel, *Mémoires pour servir à l'histoire du
Jacobinism,* 4 vols., London, 1797, v. 1, p. 22; v. 2, p. 325.

236 *Cagliostro's supposed plot:* Umberto Eco, *Serendipities: Language and Lunacy,*
New York, 1998, pp. 13–24.

236 *I may be:* Alexandre Dumas, *Memoirs of a Physician,* London, p. 536.

237 *to be launched:* Franco Riccomini, *L'enigma Cagliostro,* Florence, 1995,
p. 140.

238 *The Magic Flute:* Jacques Chailley, *The Magic Flute Unveiled: Esoteric Symbolism
in Mozart's Masonic Opera,* Rochester, Vt., 1992, pp. 59, 77, 121.

238 *for writing prophetic:* Marsha Keith Schuchard, "Blake and the Grand
Masters," pp. 6-7; William Blake, "The French Revolution. Book the

First," *Poetry and Prose of William Blake*, ed. Geoffrey Keynes, London, 1967, p. 167.

238 *anti-Inquisition works*: Philippe Jacques de Loutherbourg, *A Catalogue of All the Valuable Drawings, Sketches, Sea Views, and Studies... Also Includes Library of Scarce Books*, London, 18 June 1812; *Catalogue of Very Curious, Extensive and Valuable Library of Richard Cosway*, London, 8 June 1821.

239 *divine spiritual assistance*: Mary Pratt, *A List of a Few Cures Performed by Mr. and Mrs. de Loutherbourg at Hammersmith Terrace, with Medicine, by a Lover of the Lamb of God*, London, [1789,] pp. 1–5.

240 *driven rapidly across*: Quoted in Stephen Lloyd, *Richard and Maria Cosway: Regency Artists of Taste and Fashion*, Edinburgh, 1995, pp. 93–95.

240 *prophetic illustrations*: Morton D. Paley, *The Apocalyptic Sublime*, New Haven, Conn., and London, 1986, pp. 53–56.

240 *man looks on*: Paley, *Apocalyptic Sublime*, p. 54.

240 *oppression of the spirit*: Anderson, *The Legend of the Wandering Jew*, especially pp. 128–131; Stableford, *Tales of the Wandering Jew*, pp. 148–190; Martin Gardner, "The Wandering Jew and the Second Coming," in *The Night Is Large: Collected Essays 1938–1995*, Harmondsworth, 1996, pp. 525–532.

241 *his own legend*: Riccomini, *L'enigma Cagliostro*, pp. 135–136.

242 *being a liar*: Casanova, *Soliloque d'un penseur*, pp. 39–40.

243 *the eighteenth century*: Carlyle, "Count Cagliostro: In Two Flights" [1833], p. 301.

243 *their own greed*: Carlyle, "The Diamond Necklace" [1837], *Essays*, v. 3, pp. 321–402.

244 *not fulfilling*: Thomas Carlyle, *The French Revolution*, Oxford, 1989, p. 453.

244 *Knights Templars*: Umberto Eco, *Faith in Fakes. Travels in Hyperreality*, London, 1986, pp. 21–100; see also his "Migrazioni di Cagliostro," in *Presenza di Cagliostro*, pp. 135–151.

Notes

245 *fetishists of reason:* Jeffrey Mehlman, *Walter Benjamin for Children. An Essay on his Radio Years*, Chicago, 1993, pp. 42–60.

245 *figures of all time:* Spence, *Encyclopedia of the Occult*, p. 85.

245 *charge of her soul:* Gulley, *Encyclopedia of Mystical and Paranormal Experience*, 1991, p. 553.

245 *Egyptian theosophy:* Godwin, *Theosophical Enlightenment*, 1994, pp. 277–306.

246 *Count of Cagliostro:* Vincenzo Salerno, "Cagliostro," *Best of Sicily*, 12 June 2002, www.bestofsicily.com/mag/art15.htm